CULTURAL POLITICS

Volume 19, Issue 1
March 2023

MULTISPECIES JUSTICE

Special Issue Editors

Danielle Celermajer and Sophie Chao

Volume 19, Issue 1
March 2023

MULTISPECIES JUSTICE

Special Issue Editors
Danielle Celermajer and Sophie Chao

INTRODUCTION

Multispecies Justice

Sophie Chao and Danielle Celermajer

Abstract **This introduction to the special issue "Multispecies Justice" traces various histories and genealogies of multispecies justice, illuminating the critical contributions of Indigenous philosophies and lifeways and more recent justice movements and intellectual developments in the West. It emphasizes how these intellectual traditions are rooted in social and political movements spurred by the relentless violence against the more-than-human and the inadequacy of existing conceptualizations or institutions of justice. The introduction explains the issue's engagement with the relationship between epistemological cultures and cultural ontologies on the one hand, and political institutions on the other, with a particular focus on different "species" of beings (human, nonhuman animal, plant, and so on). It also sets out the methodological and representational challenges involved in conceptualizing and achieving multispecies justice. The introduction introduces the articles to follow by thematizing them around four key topics: the relationship between agency and representation; situated knowledges and knowledge production; colonialism and capitalism; and the law and institutions understood as formal rule-systems *and* informal rules and norms. By engaging these themes, the special issue seeks to imagine how political institutions might be formed and transformed in ways that are responsive to cultural ontologies that disrupt existing grids of meaning and distributions of value.**

Keywords **multispecies justice, more-than-human, cultural ontology, posthumanism, political institutions, representation, epistemology**

As the pressures of human exploitation of the planet intensify, the experiences of injustice among differently located humans, other-than-humans, and the environment deepen and become more glaring. It is also becoming

Cultural Politics, Volume 19, Issue 1, © 2023 Duke University Press
DOI: 10.1215/17432197-10232431

increasingly apparent that hegemonic existing frameworks for conceptualizing justice, the ways in which justice is imagined and represented, and the dominant political institutions for delivering justice are not up to the task of attending to the multiple dimensions and experiences of injustice in a multispecies world. Specifically, the individualist and humanist ontologies and ethical frameworks that underpin virtually all theories and practices of justice in Western legal cultures are proving inadequate to encompass the needs, relations, interests, concerns, communicative styles, and lifeways of more-than-human beings, and indeed many humans as members of diverse yet shared communities of life.

This special issue seeks to contribute to the larger exploration that is currently taking place across various disciplines of what it would mean to reconceptualize, reimagine, and reinstitute justice through a multispecies lens. While the term *justice* implies an emphasis on institutional protections, the very possibility of political or legal institutions commencing the task of doing justice in a multispecies world requires acknowledging the presence of beings endowed with perceptual lifeworlds, communicative capacities, biotic affordances, and ecological situatedness that may be radically different from those of humans. This, in turn, calls on an acknowledgment of the momentous (though not necessarily insurmountable) challenges entailed in entering into each other's perceptual lifeworlds—or *Umwelten* (Uexküll 1957)—and negotiating justice in ways that honor all and different points of view. In this regard, the "multi" of multispecies justice is an acknowledgment of not simply the many different types of beings who ought to be included as subjects of justice but also the multiplicity of ways of being. The transformative work of cocreating multispecies justice thus calls for more than an expansion of the rules of inclusion that constitute politics. Rather, it requires new political imaginaries that take into account the ontological diversity, relational complexity, and incommensurable forms of communication and desire, within which just arrangements and outcomes can be cocrafted. The politics of multispecies justice is, in other words, bathed in questions of culture, politics, knowledge, and communication. Further, in response to the epistemic implications of the multiplicity of ways of being now included, the "multi" of multispecies justice also implies a multiplication of disciplinary lenses and knowledge systems.

Working across a number of fields, the contributions to this special issue explore how existing dominant political institutions and approaches to justice assume and encode particular ways of knowing and modalities of being that exclude more-than-human beings from the reach of justice. They demonstrate how dominant political institutions preclude admission except under conditions that do violence to different modalities of being and knowing. More positively, the issue engages with multispecies worlds to imagine possibilities for disrupting existing grids of meaning and distributions of value so that political, social, cultural, legal, and economic systems might be formed and transformed in ways that are responsive to, and afford the possibility of, justice for more-than-human beings. It does so through four linked lenses: agency and representation; situated knowledges and knowledge production; colonialism and capitalism; and laws and institutions.

Our foundational thesis is that comprehensively reconceptualizing, reimagining, and reinstituting justice through a multispecies lens requires something more and other than simply expanding the rules of inclusion. A *transformation* in justice calls on scholars and practitioners to challenge the ontology and representation of the subject of justice and the telos of justice itself in ways that take into account the radical diversity of ways of being, the complexity of relationships, and the impossibility (even the dangers) of assuming or seeking full comprehension or transparent communication across species lines (Neimanis in this issue). Thus, as against the strong association between justice and transparency, and the assumption that justice requires translatability (if not equivalence, a sine qua non of justice refracted through economics), justice in a multispecies context must take shared and yet fundamentally different worlds as its ground.

Such an approach demands a critical engagement with themes and processes that have long been central to our understanding of cultural politics: from ethics and aesthetics, to ideologies and values, power and performance, and colonialism and capitalism. At the same time, theorizing and enacting justice in multispecies terms invites us to reimagine culture and politics themselves as always already shaped by, and shaping, the lives, relations, and socialities of *other-than-human* beings within uneven fields of power and privilege (Tsing 2014). More than this, it demands attention to the *plurality* of ways of being and knowing that animate situated more-than-human worlds, alongside the epistemic and ontological frictions, indexicalities, and incommensurabilities produced by interspecies encounters both real and imagined.

Genealogies and Implications of Multispecies Justice

The emerging field of multispecies justice has arisen in relation to longer histories and diverse genealogies of thought and practice. It draws on a range of different fields, each of which contributes to the multispecies justice conversation through a particular sets of insights, methods, and objectives. Written from a number of disciplinary perspectives and theoretical traditions, the articles in this issue engage with a number of these genealogies. We hope that they provide not only a snapshot of contemporary thinking on multispecies justice but also a view into the intellectual traditions from which this emerging field draws its ideas and orientations. At the same time, we make no claim (in this introduction or across the articles) to provide a comprehensive intellectual biography of the idea.[1] Our intention is not to synthesize diverse intellectual traditions but to render apparent the points of tension across them and indeed to show that these tensions are central in shaping multispecies justice's evolving trajectory

At the same time, we wish to make clear that many of the intellectual traditions we trace here are rooted in social and political movements spurred by the relentless violence against the more-than-human and the inadequacy of existing conceptions or institutions of justice to bring succor or protection. Concern for the well-being of and justice toward more-than-human life-forms and the environment have given rise to a range of innovative ethical frameworks and advocacy approaches, including extending legal rights to the more-than-human, such as trees, rivers, and mountains (Stone 1972; de la Cadena 2010), or apes and elephants (Cavalieri 2001; Wise 1997), and imagining the expansion of citizenship rights to nonhuman animals

(Donaldson and Kymlicka 2011; Cochrane 2018). Such approaches radicalize the character of the subject of rights and thus may, especially through the emergence and embrace of earthrights (Cullinan 2011; Gordon 2018), eventually transform how we understand the ontological structure of legal subjecthood. Nevertheless, these efforts largely remain embedded within traditional Western ontologies. Otherwise put, much of the politics of justice remains committed to a culture in which a series of bound and exclusive associations are implicit, in particular individuality and subjecthood; individuality and the capacity to make justice claims; and, albeit often implicitly, justice and the human.

As Indigenous scholars have pointed out, these bounded associations and the nature-culture split on which they are premised are alien to Indigenous cultures. While Indigenous cultures are diverse, variously constituted through complex historical interactions, and grounded in the specificities of time, place, and community (Durie 2005; Turner 2006), their customary political, legal, and moral structures share a recognition of the coconstitutive relationships between the land and its diverse human and other-than-human dwellers (Bird-David 1990; Rose 2011; Stewart-Harawira 2012; TallBear 2015; Winter in this issue). Within this relationist ethos, other-than-humans are frequently conceived and sustained as kin (and not just kind) endowed with sentience, volition, and dignity. They participate as interagentive members within a shared community of life and are bound to other life-forms—including the human—through relations of reciprocal care and nurture (Kimmerer 2013). These relations are in turn inextricable from the places in which they are embedded (Todd 2014). They also operate intergenerationally in that they recognize how just interspecies relations in the past may enable livable presents and futures, both within and across species lines (Winter 2020, 2021).

Sustained critiques of liberal humanism from Indigenous and anticolonial perspectives have more recently been joined by a number of bodies of theory and research practices in the Western academy. Posthumanist and new materialist approaches, for instance, work to reembed human beings, aims, and goals in mutually constitutive relations, or "intra-actions" with other species, elements, and technologies (Jane Bennett 2009; Braidotti 2017; Barad 2003; Haraway 2008). Cosmopolitical and ontological theorists problematize the question of who and what composes the common world, questioning the assumed singularity of reality and drawing attention to the exclusions, potentialities, and politics generated by difference and disagreement over what counts *as* and *in*, more-than-human worlds (de la Cadena 2010; Latour 2004; Stengers 2010). Actor network theorists, meanwhile, invite a flattening of ontological difference through an attention to the complex, contextual, and changing networks of people, organisms, things, and processes in ways that belie modernist classifications and their constructed but naturalized splitting of the cultural from the natural, the subject from the object, and the agentive from the structural (Latour 2005, 2017).

Critiques of liberal humanism are also central to the interdisciplinary currents of environmental humanities and multispecies studies, which aim to bridge conventional divides between the natural sciences, arts, and humanities in describing and theorizing naturecultures through the situated agencies of their human and other-than-human dwellers (Rose et al. 2012; van Dooren, Kirksey, and Münster

2016; Kirksey and Helmreich 2010). A similar project animates the work of scholars in critical animal studies (Wolfe 2003; Gruen 2015) and critical plant studies (Myers 2017; Ives 2019; Foster 2019), who analyze and often challenge the entrenched concatenations of structural, institutional, technoscientific, and discursive powers shaping how animal and vegetal life-forms are classified, hierarchized, and instrumentalized to serve (certain) human ends.

The currents outlined above, while certainly not exhaustive, offer capacious and comprehensive ways to critique humanist and individualist ontologies and reconceptualize the beings who might be legitimate subjects of concern or interest. They do so by resituating the human within a broad spectrum of life in which other-than-human organisms, once relegated to the status of bare life, or *zoe*, are repositioned as (co)makers of meaning endowed with fleshy cultural, historical, and political biographies and involved in sympoetic becomings. Still more radically, some of these theoretical approaches decenter the individual subject altogether and shift ontological primacy to material-semiotic relationships, ecologies, assemblages, processes, and so forth (Barad 2012; Haraway 2008; Kohn 2013).

Just as important to the enterprise of multispecies justice are the critiques and contestations that have arisen in response to these various approaches, all of which must inform how justice beyond the human is theorized across and beyond disciplinary imaginaries. These critiques include worries about a lack of attention to the uneven power dynamics shaping assemblages of beings and things within actor network theory; the dangerous slippage of posthumanism toward anti- or ahumanism; the elision of thought, language, and discourse as sources of meaning and world-making within rigidly materialistic approaches; and the challenges for cosmopolitical and ontological approaches of achieving translation or interaction *between* worlds. Other critiques call out the exclusion or unacknowledged appropriation of Indigenous knowledges within Western theoretical currents; the frequent neglect of questions of race, gender, and (dis)ability in shaping human and other-than-human lifeworlds; and the erasure of colonialism and its afterlives in the constitution of what counts as knowledge, science, philosophy, and theory, and who gets to produce and use it.

Even as the move beyond the human and associated intellectual debates seem to promise a more capacious and nuanced understanding of justice, it remains critical to be vigilant against eliding the strategies of dehumanization that have organized systematic intrahuman injustice and violence and critiques of the multiple forms of violence generated by humanism. Such vigilance demands attending to the morally laden and racialized instrumentalization of "species" categories and hierarchies by dominant human groups in order to legitimate the exploitation of peoples as fungible bodies, extractible labor, dangerous vermin, and disposable property (Moore, Pandian, and Kosek 2003; Hage 2017; Joshua Bennett 2020; King 2019; Mavhunga 2011). Navigating the cultural politics of justice beyond the human, in other words, demands reflexive consideration of the ways in which colonial-capitalist-racist assemblages and their afterlives undermine the flourishing of multispecies communities of life *and* continue to relegate certain human populations to the status of subhuman, nonhuman, and killable before the law (Büscher 2022; Gilroy 2017; Jackson 2015; Kim 2015; Murphy 2017; Weheliye 2014). Fields such as ecofeminism (Plumwood 1998; Mies and Shiva 1993), postcolonial studies (Ahuja 2009; Caluya 2014), critical race studies (Joshua Bennett 2020; Boisseron 2018; Jackson 2020; Wynter 2006), and critical disability studies (Chen 2012; Clare 2015; Puar 2017) insist that any "turn" to the nonhuman must be crucially informed by counterhegemonic understandings of justice that call into question the exclusions encoded into the assumed Western liberal subject of rights. The capacious justice we seek must, then, also be grounded in such perspectives.

Until recently, attention to justice in multispecies terms was relatively fragmented, confined to particular subfields, or embryonic (Alaimo 2019; Haraway 2018; Heise 2016; Kirksey 2017; Radomska 2017). A growing body of interdisciplinary scholarship is now seeking to more systematically reconceptualize justice beyond the human realm and draw out the implications of this reconceptualization for the ideological and structural transformation of existing political institutions. Without seeking to summarize a field that is growing as we write, a number of themes are emerging. A nonexhaustive list includes: the disruption of structural anthropocentrism and nature/culture divides and the invitation to embrace multispecies justice as a critical path to better shared future worlds (Celermajer et al. 2020; Thaler 2021); the emergence of interspecies responsibility through encounters with other-than-human beings in the face of climate change–induced and unevenly distributed vulnerabilities (Tschakert 2020; Tschakert et al. 2021); multispecies justice praxis grounded in acts of multispecies love—affective, practical, and political—enacted across diverse, interconnected communities of life (Fernando 2020); possibilities of justice approached through the lens of contingency, situatedness, and partial connections (Chao 2021a; Chao, Bolender, and Kirksey 2022; Heise and Christensen 2020; Weaver 2021); the importance of artistic production and visual cultures in a world increasingly shaped by climate breakdown (Agarwal 2021; Broglio 2021; Celermajer et al. 2020; Demos, Scott, and Banerjee 2021a); and the centrality of capitalism and colonialism to multispecies injustice (Chao 2021b, 2022; Celermajer 2020; Celermajer et al. 2021; Emel and Nirmal 2021; Gillespie and Collard 2015).

This special issue seeks to take up some of the openings suggested in this emerging field, in ways that will expand and deepen possibilities for practices of living together that are hospitable to a broader range of subjects at a time of socioecological unraveling, threat, and instability. The articles reflect on what this transformation in justice might mean for human lifeworlds and their inextricable yet always historically and culturally situated relationship with the more-than-human. Rather than simply celebrating the fact of more-than-human mingling, we follow Donna Haraway in honestly asking what a more responsible "sharing of suffering" across species lines might look like in forging "barely possible but absolutely necessary joint futures" (Haraway 2008: 72, 2003: 7).

Pushing against anthropocentric, hierarchist, and individualist understandings of justice, and correlatively of knowledge, representation, culture, politics, and sociality, we ask: What does justice mean when refracted through a multispecies lens? Who/what is justice for and who benefits from justice? How ought we conceive of the subject of justice? Does it even make sense to speak of subjects of justice in multispecies worlds? What is the relationship between justice and other areas or dimensions of ethics such as care or hope? How do different practices of research, knowledge, representation and communication impede or enable the possibility of justice across radical difference?

Within this potentially vast intellectual space, the particular focus of this issue is on the relationship between epistemological cultures and cultural ontologies on the one hand and political institutions on the other, with a particular focus on different "species" of beings (human, nonhuman animal, plant, and so on). By *epistemological cultures*, we refer to the different ways in which people's capacities to know, act toward, and form relationships (including relationships of justice) with different types of beings are shaped; how different types of beings come to matter for differently located humans, and specifically, the ethical stakes of these different ways of knowing and relating. The complementary term, *cultural ontologies*, refers to the quality or character of the being of different beings and their status within meaning-laden grids. Following Bruno Latour (2004), we pair epistemological cultures and cultural ontologies to underline that recognizing the radical differences across cultures requires acknowledging that there is not one cosmos that transcends different "local cultures" but differently composed cosmoses. Understanding what justice might entail cannot be achieved without interrogating these epistemological and ontological premises (Ruiz-Serna, in this issue). Indeed, taking ontological politics a step further and in a multispecies direction would demand wondering about the cultures and ethical orientations of beings other than humans. In this regard, and recognizing the inextricable entanglement of *bios* and *geos*, the special issue also engages with elements that might normally be excluded even from a multispecies geography—for instance, oceans, soils, and territories (Povinelli 2016; TallBear 2015; Todd 2017; Reid in this issue).

Thus understood, we are interested in how existing dominant political institutions encode epistemological cultures and cultural ontologies in ways that exclude beings other than humans from the category of subjects of justice and indeed condition and sanction systematic violence against them (Singer in this issue). At a more foundational level, these institutions preclude admission except under conditions that do violence to different epistemological cultures and cultural ontologies. Thus, while the articles in the issue are interested in recent developments in the recognition of nonhuman legal personhood as well as precedents in Indigenous and other non-Western ontologies, they caution against prioritizing (anthropogenic) juridical spaces and instruments in ways that render opaque the potentials afforded by nonjuridical, "social," artistic, and more-than-human practices and phenomena that may prove more conducive to radical innovation. In this regard, we adopt a generous understanding of the means of and to justice.

The special issue aims to imagine how political institutions might be formed and transformed to become responsive to cultural ontologies that disrupt existing grids of meaning and distributions of value. It does this in two principal ways. First, it considers what happens to the idea and practice of multispecies justice if we take as a starting point epistemological cultures and cultural ontologies that radically challenge those that are assumed by and inform dominant justice institutions. Second, it considers what happens to the idea and practice of multispecies justice if one assumes as one's starting point the being of beings other than humans (Nassar and Barbour in this issue). In other words, we interrogate what beings other than humans might suggest to humans about justice and the politics of life, not only in a naturalistic sense (i.e., how they are in some idealized context) but also in contexts of past and ongoing colonialism, nationalism, and capitalism (Chatterjee in this issue).

This special issue also explicitly aims to explore the methodological and representational challenges involved in conceptualizing and achieving multispecies justice. If the pursuit or fantasy of full translation and transparent communication are features of an anthropocentric conception of justice (and even then, a narrow and excluding one), multispecies justice invites us to consciously and explicitly experiment with a different set of tools, processes, and objectives. We are interested, for example, in ideas like care (including complicit care), compromise, respectful distance, and imagination. Following Anna Tsing's (n.d.) invitation to experiment with disruptive grammars in the "big human mess" that is the current ecological epoch, we seek to develop a (re)new(ed) set of more-than-human vocabularies and representational practices that can better capture the contradictions, dilemmas, and hopeful openings at play in achieving justice across and within multispecies worlds.

Fulfilling, or even approaching the fulfilment of multispecies justice cannot be achieved within the boundaries of any single discipline, nor even within the boundaries of the traditional academy, nor within the institutional borders that cordon off scholars, artists, and activists (Demos, Scott, and Banerjee 2021a). For this reason, this special issue includes contributions from scholars across a range of humanities and social science disciplines, collaborative work between humanities scholars and natural scientists, and contributions from artists and activists. Articles within the issue engage with methods and concepts derived from fields including cultural theory, anthropology, political theory, philosophy, art, history of science, queer/feminist theory, Indigenous studies, law, conservation science, and plant science. Included in this issue also are works by Ravi Agarwal, Janet Laurence, and David Brook, whose artistic endeavors sit at the interface of scholarship and advocacy. Fostering diverse modes of engagement with theory, cultural production, and politics, these artistic contributions independently convey and perform themes concerning multispecies justice rather than illustrating the ideas suggested in the essays. They are critical to performing the methodological innovation we believe is required to explore if not enact multispecies justice.

One of our hopes in curating this special issue is that it will engage scholars who do not already identify as working on multispecies or indeed animal or environmental issues at all, but whose concerns are profoundly relevant to and intersect with ours on conceptual or political dimensions. We are interested in conversing with scholars who are exploring the intersections between ideas and practices concerning identity, culture, politics, law, power, representation, and modes of communication. Insofar as articles in this special issue take up these topic areas in the context of questions about justice for beings other than humans, and/or humans and more-than-human beings in relationship, they promise to offer fresh perspectives on the topics outlined above. We see our collection as offering something new to engagements with questions of identity-based power, inequality, and marginalization that have not yet comprehensively or programmatically taken up the multispecies question.

In the final section of this introduction, we outline a number of recurring themes that animate our interdisciplinary foray into the realm of multispecies justice: agency and representation, situated knowledges and knowledge production, colonialism and capitalism, and laws and institutions. We hope our readers will see our engagements with these themes as an invitation to identify theoretical, conceptual, or empirical affinities with their scholarship in ways that will catalyze innovation in our mutual work and across disciplines.

Themes and Articles

While diverse in their thematic and theoretical scope, the empirical and conceptual contributions of this issue revolve around a number of key interrelated issues. First is the relationship between agency and representation. Continental philosopher Dalia Nassar and plant physiologist Margaret Barbour take up the question of agency in relation to trees, a community of species that have, until relatively recently, been excluded from ethical, moral, and political purview in Western thought. Synthesizing their respective insights from philosophy and plant science, Nassar and Barbour offer the concept of "embodied history" to reframe trees as inherently relational beings that hold in their very materiality the biological, historical, geological, and ecological processes that together produce life and the environment. Approaching trees as embodied history invites us to reckon with vegetal beings as ethical subjects beyond the realm of representation. It also brings into question speciesist hierarchies that have been encoded in justice theories and that obscure forms of historical and ecological agency stemming from outside zoocentric and anthropocentric notions of sentience and intelligence. In this regard, their article asks not what justice might do for trees but what trees might do for justice.

Agency and representation meet vulnerability, materiality, the law, and capitalist extraction in cultural theorist Susan Reid's essay on ocean justice. Thinking-with oceans through the lens of mastery, discursivity, alterity, and imagination, Reid articulates vulnerability as an agentic force that, if acknowledged and accommodated by institutions, can reveal how exposures to harm extend in networks that vastly transcend individual human subjects to include interdependent organismic and elemental actors. A vulnerability-based approach thus opens generative pathways toward imagining legislative and economic institutions that could, contra the dominant regime of extractive capitalism and so-called ocean justice, offer the possibility of securing the materials humans need to live well while also ensuring the adequate recognition of other-than-human beings and the flourishing of ocean worlds.

Anthropologist Daniel Ruiz-Serna's essay, meanwhile, foregrounds how Indigenous and Afro-Colombian peoples conceptualize and relate to "territory" not just as physical lands but rather as sets of emplaced and agentive relationships through which humans share life with much wider assemblages of human and other-than-human beings—venomous snakes, guardian spirits, monocrop oil palms, and sentient forests, among others. These conceptualizations challenge conventional paradigms of politics in general and transitional justice in particular as a human-only activity and of multiculturalism as a representational elision of ontological difference. Instead, they draw attention to the existential stakes of ecological destruction for the many worlds and world-making practices that together produce the territory as a multiplicitous, agential meshwork of matter and meaning.

Agency and representation resurface in art historian and environmental humanities scholar Sria Chatterjee's essay on art, design, and plant sentience. Tracing the co-optation of vegetal sentience beyond scientific discourse into cultural and ideological fields through artistic representations, Chatterjee demonstrates how efforts to rehabilitate vegetal agency remain mediated at the core by a range of anthropocentric discourses and historically engrained relationships with colonial, nationalist, and capitalist world systems. Calling into question the assumption that more inclusion means more justice, her essay demonstrates how these systems and discourses profoundly shape whether the recognition of plant agency can lead to just ethical or political outcomes for plants themselves, or whether such recognition is simply circled back to serve particular human needs and hegemonic ideologies.

Multispecies violence—both real and representational—takes center stage in ecofeminist theorist and critical animal studies scholar Hayley Singer's essay, which meditates on the factory farm as a form of hell for animals whose horrific lives and deaths are normalized under institutionalized agro-industrial regimes. Singer's contribution grapples with the complexities of articulating and enacting multispecies literary justice through textually grounded modes of empathy, subjectivity, and point of view in relation to beings other than human. Her approach attends to the visceral granularity of industrial and flesh realisms in order to broaden representations of multispecies life and death and to foster solidarities across species differences, similarities, and complexities. Challenging the form of the essay itself, Singer offers a mode of noninnocent representative thinking, committed to the difficult but necessary political and ethical labor of dissolving hegemonic and humancentric subject-writer-reader distinctions.

Imagining and enacting multispecies justice draws attention to the broader power relations that shape particular processes of knowledge and value production and their more-than-human consequences (Haraway 1988). Situated knowledges and knowledge production thus constitutes a second recurring motif in the contributions to this special issue. Returning to the Old Norse etymology of *hell* (*hel*, or "cover"), Singer examines the narrative strategies through which literary texts keep secret, hidden, or covered over, the unspeakable horrors of animal cruelty in Concentrated Animal Feeding Operations (CAFOs) and factory farms, such that they cannot be adequately apprehended by the human mind. A practice of "poetic outrageousness," Singer argues, can offer more ethically rigorous representations of the acts of intolerable violence that govern animal life and death in industrial settings.

Chatterjee, meanwhile, traces the colonial and aesthetic mobilization of scientific knowledge about vegetal life-forms from late nineteenth-century Indian political spheres to contemporary Western neurobiological and bioengineering settings. In particular, she examines how the Western knowledge system of plant science was instrumentally activated to further human causes as diverse as Hindu nationalist ideology in India and biomimetic technological innovations in modern capitalist and military-industrial frameworks.

Reid takes up the question of knowledge production and its limits in relation to the United Nations Convention on the Law of the Sea, whose primary focus on "ocean development" renders it incapable of adequately responding to the direct harms of extraction. Indeed, she points to its complicity with them. Drafted in the 1970s under the pressures of nation-states and powerful industry lobbyists, the Convention conjures the ocean into territorial zones of exploitation that bear no resemblance to the actual ocean, thereby producing and compounding the ocean's material vulnerability to now naturalized modes of anthropogenic exploitation.

The violence of knowledge production as a form of extraction also constitutes a central thematic in culture and gender theorist Astrida Neimanis's contribution to this special issue. Neimanis examines how groundwater-dwelling stygofauna—a deep-time invertebrate species—unsettle the assumption that knowledge, care, and justice must necessarily be predicated on a symbolic and literal practice of "revelatory" knowing. Stygofauna's resistance to ocularcentric Western epistemologies, together with their growing vulnerability to industrial coal mining, troubles the ethical implications of knowing in the name of justice when it is tangled up with knowing as further violence. Bringing Indigenous water management practices into conversation with narrative fiction writing, Neimanis identifies other ways of knowing that take nonhuman strangeness and opacity as the basis for cultivating new kinds of ethical relations, forged through childhood memories, stories, embodied encounters, sounds, Indigenous law, conversation, and various kinds of science.

Extractive modes of knowing are further challenged in political theorist Christine Winter's essay, which examines how Mātauranga Māori—the epistemological foundations of Māori philosophy and science—is both generated by and generative of, layers of living, nonliving, ancestral, and spiritual beings bound in interdependent relationships (*whakapapa*). Recognizing and respecting the ongoing significance and efficacy of Indigenous knowledge, science, philosophy, and culture in nurturing multispecies relationships and well-being, and engaging with Indigenous peoples as intellectual peers and producers of knowledge, Winter argues, is of fundamental importance if multispecies justice is to counter both the damaging domination of the nonhuman realm *and* the ongoing colonial domination of Indigenous epistemologies and ontologies in and beyond the academic sphere. Such an approach would ground itself in the principle of relationality, covering all planetary being within an expansive nonmechanical, nonlinear conception of time/space/matter that is animated by multiple, more-than-human epistemologies and ontologies.

Envisioning justice beyond the human demands attention to the perduring afterlives of colonial regimes, as these manifest in the ongoing epistemic and material violence wrought by agro-industrial, capitalist landscape transformations and attendant forms of ecological degradation. Colonialism and capitalism thus constitute a third recurrent theme in this issue. Analyzing the rise of biomimesis as a dominant techno-natural fix for the future, Chatterjee calls for the development of a framework of politics and ethics that departs from framing the natural world as dependent on human interference and innovation. This framing, she points out, problematically erases inequalities (between humans and other sentient beings) and colonial structures of power and domination even as they persist in current geopolitical and economic structures. Through a comparative analysis of Gaganendranath Tagore's 1921 satirical picture "Reform Screams" and contemporary artist Pedro Neves Marques's video "The Pudic Relation between Machine and Plant," Chatterjee also reflects on the conflation of human and plants as colonial subjects and as vehicles for political nationalism within entangled histories of imperial and economic domination.

Neimanis, meanwhile, describes how quests for interspecies intimacy within colonial scientific knowledge projects are often aligned with colonial extractivist projects. She invites decolonial, eco-crip, and poetic ways of knowing that depart from scientistic positivism and find roots in the deeply embodied and situated nature of nature.

Reid's essay, on the other hand, performs a decolonial analytical move by thinking-with "multibeing justice" rather than "multispecies justice." Such a reframing, she posits, resists the violent logics of colonialism and biological determinism embedded within the Linnaean term *species*—one that elides the relational being and becomings of both organismic and elemental entities.

Writing as a citizen of Aotearoa New Zealand and Australia, where Indigenous peoples' sovereign claims remain insufficiently recognized by the contemporary settler colonial state, Winter cautions against the risk of academics perpetuating colonial knowledge practices through hegemonic and exclusionary modes of theory and practice. Eschewing the universalist impulse and instead embracing the powerful potential of multiple philosophical traditions, Winter invites a decolonial approach to multispecies justice, where justice is done not to individuals or species but rather to spatiotemporally expansive sets of relationships whose scale of mattering is open and accountable to all there is: human, animal, vegetable, mineral, and spiritual. Such an approach also challenges the neoliberal capitalist ethos of settler colonial regimes by reimagining wealth as that which comes from caring for and carefully tending to the health and wellbeing of interdependent human, nonhuman, and spiritual realms.

As Winter's essay illustrates, emergent theorizations of multispecies justice within the Western academy rub up against strongly held notions about who counts ethically and politically before the law. The law and institutions thus represent the fourth and final central theme in this issue, with "institutions" defined in the broadest sense to encompass both formal rule systems (judicial systems and legislation, constitutions and political bodies, economic systems) *and* informal rules and norms (species, personhood, gender, race) that coordinate, discipline, and manage actions (Celermajer, Churcher, and Gatens 2020).

Ruiz-Serna explores the ontological and epistemological frictions between Indigenous, Afro-Colombian, and state understandings of nature in the context of recent legislative measures that recognize traditional territories as victims of war in postconflict Colombia. This national legal precedent, Ruiz-Serna argues, foregrounds an important shift from notions of "territorial damage," or actions that limit the effective enjoyment of ownership rights, to notions of "damage to territory," or actions that jeopardize the relationships that communities cultivate with the myriad beings who constitute their territories. In the process, new possibilities arise for mending wrongs of war that go beyond considerations of human or environmental damage and that offer a unique opportunity to decolonize justice and decenter the human in our understandings of war and its aftermath.

Reid takes up jurisdictional institutions in her propositional analysis of responsible cohabitation with watery worlds. Legal and scientific paradigms, Reid demonstrates, routinely position the ocean and its more-than-human dwellers as economic resources destined for human extraction and exploitation rather than as lively ecological actors and potential subjects of justice. Symptomatic of Western, anthropocentric ideologies of mastery and control, these institutions patrol the limits of who and what counts before the law and claim exclusive authority over ways of knowing the ocean. Achieving multispecies justice in an oceanic context, Reid argues, is not exclusively a matter of needing better law or more marine scientific knowledge but rather of learning how to perceive and relate to the ocean in intersubjective, and not exploitative, terms.

Indigenous protocols offer a vital avenue for reclaiming the law as an instrument for the fulfilment of multispecies justice. Neimanis, for instance, examines how Customary First Law in Aboriginal Australia has guided the responsible and respectful management and care of lands and waters since time immemorial, including in the absence of empirical and accessible forms of evidence for why this care matters. Winter, meanwhile, offers an important corrective to the framing of the academic field of multispecies justice as "new" by examining its long-standing existence as a field of philosophy, protocol, and practice among Indigenous peoples, for whom Western-derived nature-culture binaries are anathema to relational living and thinking. As such, the recognition of nonhuman entities as legal persons with interests, rights, powers, and duties, too, Winter argues, must recognize the shared identity and belonging of both nature and its human custodians, as these are produced through cascading spirals of time and relation.

As we hope is evident from this overview, our four principal themes—agency and representation, situated knowledges and knowledge production, colonialism and capitalism, and laws and institutions—are entwined through the essays in this special issue in ways that demonstrate the weave of culture and politics that we set out at the beginning of this introduction. We trust that readers, coming to our collective reflections with their own perspectives, concerns, and experiences, and those of the multispecies worlds in which they are embedded, will draw their own threads beyond this offering.

Note

1. For a more extended but also noncomprehensive genealogy of the contemporary work on multispecies justice, see Celermajer et al. 2021.

References

Agarwal, Ravi. 2021. "Alien Waters." In Demos, Scott, and Banerjee 2021b: 365–75.

Ahuja, Neel. 2009. "Postcolonial Critique in a Multispecies World." *PMLA* 124, no. 2: 556–63.

Alaimo, Stacy. 2019. "Afterword: Crossing Time, Space, and Species." *Environmental Humanities* 11, no. 1: 239–41.

Barad, Karen. 2003. "Posthumanist Performativity: Toward an Understanding of How Matter Comes to Matter." *Signs* 28, no. 3: 801–31.

Barad, Karen. 2012. "On Touching—The Inhuman That Therefore I Am." *differences* 23, no. 3: 206–23.

Bennett, Jane. 2009. *Vibrant Matter: A Political Ecology of Things*. Durham, NC: Duke University Press.

Bennett, Joshua. 2020. *Being Property Once Myself: Blackness and the End of Man*. Cambridge, MA: Harvard University Press.

Bird-David, Nurit. 1990. "The Giving Environment: Another Perspective on the Economic System of Gatherer-Hunters." *Current Anthropology* 31, no. 2: 189–96.

Boisseron, Bénédicte. 2018. *Afro-Dog: Blackness and the Animal Question*. New York: Columbia University Press.

Braidotti, Rosi. 2017. "Critical Posthuman Knowledges." *South Atlantic Quarterly* 116, no. 1: 83–96.

Broglio, Ron. 2021. "Multispecies Futures through Art." In Demos, Scott, and Banerjee 2021b: 342–52.

Büscher, Bram. 2022. "The Nonhuman Turn: Critical Reflections on Alienation, Entanglement, and Nature under Capitalism." *Dialogues in Human Geography* 12, no. 1: 54–73. https://doi.org/10.1177/20438206211026200.

Caluya, Gilbert. 2014. "Fragments for a Postcolonial Critique of the Anthropocene: Invasion Biology and Environmental Security." In *Rethinking Invasion Ecologies from the Environmental Humanities*, edited by Jodi Frawley and Iain McCalman, 31–44. London: Routledge.

Cavalieri, Paola. 2001. *The Animal Question: Why Nonhuman Animals Deserve Human Rights*. Translated by Catherine Woollard. Oxford: Oxford University Press.

Celermajer, Danielle. 2020. "Rethinking Rewilding through Multispecies Justice." *Animal Sentience*, no. 28. https://www.wellbeingintlstudiesrepository.org/animsent/vol5/iss28/12/.

Celermajer, Danielle, Sria Chatterjee, Alasdair Cochrane, Stefanie Fishel, Astrida Neimanis, Anne O'Brien, Susan Reid, Krithika Srinivasan, David Schlosberg, and Anik Waldow. 2020. "Justice through a Multispecies Lens." *Contemporary Political Theory* 19, no. 3: 475–512.

Celermajer, Danielle, Millicent Churcher, and Moira Gatens. 2020. *Institutional Transformations: Imagination, Embodiment, and Affect*. New York: Routledge.

Celermajer, Danielle, David Schlosberg, Lauren Rickards, Makere Stewart-Harawira, Mathias Thaler, Petra Tschakert, Blanche Verlie, and Christine Winter. 2021. "Multispecies Justice: Theories, Challenges, and a Research Agenda for Environmental Politics." *Environmental Politics* 30, nos. 1–2: 119–40.

Chao, Sophie. 2021a. "The Beetle or the Bug? Multispecies Politics in a West Papuan Oil Palm Plantation." *American Anthropologist* 123, no. 2: 476–89.

Chao, Sophie. 2021b. "Can There Be Justice Here? Indigenous Experiences in the West Papuan Plantationocene." *Borderlands* 20, no. 1: 11–48.

Chao, Sophie. 2022. *In the Shadow of the Palms: More-Than-Human Becomings in West Papua*. Durham, NC: Duke University Press.

Chao, Sophie, Karin Bolender, and Eben Kirksey, eds. 2022. *The Promise of Multispecies Justice*. Durham, NC: Duke University Press.

Chen, Mel Y. 2012. *Animacies: Biopolitics, Racial Mattering, and Queer Affect*. Durham, NC: Duke University Press.

Clare, Eli, ed. 2015. *Exile and Pride Disability, Queerness, and Liberation*. Durham, NC: Duke University Press.

Cochrane, Alasdair. 2018. *Sentientist Politics: A Theory of Global Inter-species Justice*. Oxford: Oxford University Press.

Cullinan, Cormac. 2011. *Wild Law: A Manifesto for Earth Justice*. 2nd ed. White River Junction, VT: Chelsea Green.

de la Cadena, Marisol. 2010. "Indigenous Cosmopolitics in the Andes: Conceptual Reflections beyond 'Politics.'" *Cultural Anthropology* 25, no. 2: 334–70.

Demos, T. J., Emily E. Scott, and Subankhar Banerjee. 2021a. Introduction to Demos, Scott, and Banerjee 2021b: 1–10.

Demos, T. J., Emily E. Scott, and Subankhar Banerjee, eds. 2021b. *The Routledge Companion to Contemporary Art, Visual Culture, and Climate Change*. New York: Routledge.

Donaldson, Sue, and Will Kymlicka. 2011. *Zoopolis*. Oxford: Oxford University Press.

Durie, Mason. 2005. *Nga Tai Matatu: Tides of Maori Endurance*. Melbourne: Oxford University Press.

Emel, Jody, and Padini Nirmal. 2021. "A Feminist Research Agenda for Multispecies Justice." In Hovorka, McCubbin, and Van Patter 2021: 23–37.

Fernando, Jude L. 2020. "From the Virocene to the Lovecene Epoch: Multispecies Justice as Critical Praxis for Virocene Disruptions and Vulnerabilities." *Journal of Political Ecology* 27, no. 1: 685–731. https://doi.org/10.2458/v27i1.23816.

Foster, Laura A. 2019. "Critical Perspectives on Plants, Race, and Colonialism: An Introduction." *Catalyst: Feminism, Theory, Technoscience* 5, no. 2: 1–6.

Gillespie, Kathryn, and Rosemary-Claire Collard, eds. 2015. *Critical Animal Geographies: Politics, Intersections, and Hierarchies in a Multispecies World*. New York: Routledge.

Gilroy, Paul. 2017. "'Where Every Breeze Speaks of Courage and Liberty': Offshore Humanism and Marine Xenology; or, Racism and the Problem of Critique at Sea Level." *Antipode* 50, no. 1: 3–22.

Gordon, Gwendolyn. 2018. "Environmental Personhood." *Columbia Journal of Environmental Law* 43, no. 1: 201.

Gruen, Lori. 2015. *Entangled Empathy: An Alternative Ethic for Our Relationships with Animals*. New York: Lantern.

Hage, Ghassan. 2017. *Is Racism an Environmental Threat?* Malden, MA: Polity.

Haraway, Donna J. 1988. "Situated Knowledges: The Science Question in Feminism and the Privilege of Partial Perspective." *Feminist Studies* 14, no. 3: 575–99.

Haraway, Donna J. 2003. *The Companion Species Manifesto: Dogs, People, and Significant Otherness*. Chicago: Prickly Paradigm.

Haraway, Donna J. 2008. *When Species Meet*. Minneapolis: University of Minnesota Press.

Haraway, Donna J. 2016. *Staying with the Trouble: Making Kin in the Chthulucene*. Durham, NC: Duke University Press, 2016.

Haraway, Donna J. 2018. "Making Kin in the Chthulucene: Reproducing Multispecies Justice." In *Making Kin Not Population*, by Adele E. Clarke and Donna Haraway, 67–100. Chicago: Prickly Paradigm.

Heise, Ursula. 2016. *Imagining Extinction: The Cultural Meanings of Endangered Species*. Chicago: University of Chicago Press.

Heise, Ursula, and Jon Christensen. 2020. "Multispecies Justice in the Wetlands." *European Journal of Literature, Culture, and Environment* 11, no. 2: 169–77.

Hovorka, Alice, Sandra McCubbin, and Lauren Van Patter, eds. 2021. *A Research Agenda for Animal Geographies*. Cheltenham, UK: Edward Elgar.

Ives, Sarah. 2019. "'More-than-Human' and 'Less-than-Human': Race, Botany, and the Challenge of Multispecies Ethnography." *Catalyst: Feminism, Theory, Technoscience* 5, no. 2: 1–5.

Jackson, Zakiyyah I. 2015. "Outer Worlds: The Persistence of Race in Movement 'Beyond the Human.'" *GLQ* 21, nos. 2–3: 215–18.

Jackson, Zakiyyah I. 2020. *Becoming Human: Matter and Meaning in an Antiblack World*. New York: New York University Press.

Kim, Claire Jean. 2015. *Dangerous Crossings*. Cambridge: Cambridge University Press.

Kimmerer, Robin Wall. 2013. *Braiding Sweetgrass: Indigenous Wisdom, Scientific Knowledge, and the Teachings of Plants*. Minneapolis: Milkweed Editions.

King, Tiffany L. 2019. *The Black Shoals: Offshore Formations of Black and Native Studies*. Durham, NC: Duke University Press.

Kirksey, Eben S. 2017. "Lively Multispecies Communities, Deadly Racial Assemblages, and the Promise of Justice." *South Atlantic Quarterly* 116, no. 1: 195–206.

Kirksey, Eben S., and Stefan Helmreich. 2010. "The Emergence of Multispecies Ethnography." *Cultural Anthropology* 25, no. 4: 545–76.

Kohn, Eduardo. 2013. *How Forests Think: Toward an Anthropology beyond the Human*. Berkeley: University of California Press.

Latour, Bruno. 2004. "Whose Cosmos, Which Cosmopolitics? Comments on the Peace Terms of Ulrich Beck." *Common Knowledge* 10, no. 3: 450–62.

Latour, Bruno. 2005. *Reassembling the Social: An Introduction to Actor-Network-Theory*. New York: Oxford University Press.

Latour, Bruno. 2017. *Facing Gaia: Eight Lectures on the New Climatic Regime*. Translated by Catherine Porter. Cambridge: Polity.

Mavhunga, Clapperton C. 2011. "Vermin Beings: On Pestiferous Animals and Human Game." *Social Text*, no. 106: 151–76.

Mies, Maria, and Vandana Shiva. 1993. *Ecofeminism*. London: Zed.

Moore, Donald S., Anand Pandian, and Jake Kosek. 2003. "The Cultural Politics of Race and Nature: Terrains of Power and Practice." In *Race, Nature, and the Politics of Difference*, edited by Donald Moore, Jake Kosek, and Anand Pandian, 1–70. Durham, NC: Duke University Press.

Murphy, Michelle. 2017. *The Economization of Life*. Durham, NC: Duke University Press.

Myers, Natasha. 2017. "Ungrid-able Ecologies: Decolonizing the Ecological Sensorium in a Ten-Thousand-Year-Old NaturalCultural Happening." *Catalyst: Feminism, Theory, Technoscience* 3, no. 2: 1–24.

Plumwood, Val. 1998. "Inequality, Ecojustice, and Ecological Rationality." *Social Philosophy Today*, no. 13: 75–114.

Povinelli, Elizabeth A. 2016. *Geontologies: A Requiem to Late Liberalism*. Durham, NC: Duke University Press.

Puar, Jasbir K. 2017. *The Right to Maim: Debility, Capacity, Disability*. Durham, NC: Duke University Press.

Radomska, Marietta. 2017. "The Anthropocene, Practices of Storytelling, and Multispecies Justice." *Angelaki* 22, no. 2: 257–61.

Rose, Deborah Bird. 2011. *Wild Dog Dreaming: Love and Extinction*. Charlottesville: University of Virginia Press.

Rose, Deborah Bird, Thom van Dooren, Matthew Chrulew, Stuart Cooke, Matthew Kearnes, and Emily O'Gorman. 2012. "Thinking through the Environment, Unsettling the Humanities." *Environmental Humanities* 1, no. 1: 1–5.

Stengers, Isabelle. 2010. *Cosmopolitics*. Vol. 1. Translated by Roberto Bononno. Minneapolis: University of Minnesota Press.

Stewart-Harawira, Makere. 2012. "Returning the Sacred: Indigenous Ontologies in Perilous Times." In *Radical Human Ecology: Intercultural and Indigenous Approaches*, edited by Lewis Williams, Rose Roberts, and Alastair McIntosh, 73–88. Burlington, VT: Ashgate.

Stone, Christopher D. 1972. "Should Trees Have Standing? Toward Legal Rights for Natural Objects." *Southern California Law Review* 45, no. 2: 450–501.

TallBear, Kim. 2015. "An Indigenous Reflection on Working beyond the Human/Not Human." *GLQ* 21, nos. 2–3: 230–35.

Thaler, Mathias. 2021. "What If: Multispecies Justice as the Expression of Utopian Desire." *Environmental Politics* 31, no. 2: 258–76. https://doi.org/10.1080/09644016.2021.1899683.

Todd, Zoe. 2017. "Fish, Kin, and Hope: Tending to Water Violations in Amiskwaciwâskahikan and Treaty Six Territory." *Afterall*, no. 43: 102–7.

Tschakert, Petra. 2020. "More-than-Human Solidarity and Multispecies Justice in the Climate Crisis." *Environmental Politics* 31, no. 2: 277–96. https://doi.org/10.1080/09644016.2020.1853448.

Tschakert, Petra, David Schlosberg, Danielle Celermajer, Lauren Rickards, Christine Winter, Mathias Thaler, Makere Stewart-Harawira, and Blanche Verlie. 2021. "Multispecies Justice: Climate-Just Futures with, for, and beyond Humans." *WIREs Climate Change* 12, no. 2: e699.

Tsing, Anna L. 2014. "More-than-Human Sociality: A Call for Critical Description." In *Anthropology and Nature*, edited by Kirsten Hastrup, 27–42. New York: Routledge.

Tsing, Anna L. n.d. "Catachresis for the Anthropocene: Three Papers on Productive Misplacements." In *AURA's Openings*, 2–10. Vol. 1 of *More than Human: AURA Working Papers*. Højbjerg, Denmark: Department of Culture and Society, Aahrus University.

Turner, Dale. 2006. *This Is Not a Peace Pipe: Towards a Critical Indigenous Philosophy*. Toronto: University of Toronto Press.

Uexküll, Jacob von. 1957. "A Stroll through the Worlds of Animals and Men." In *Instinctive Behavior: The Development of a Modern Concept*, edited and translated by Claire H. Schiller, 5–80. New York: International Universities Press.

van Dooren, Thom, Eben S. Kirksey, and Ursula Münster. 2016. "Multispecies Studies: Cultivating Arts of Attentiveness." *Environmental Humanities* 8, no. 1: 1–23.

Weaver, Harlan. 2021. *Bad Dog: Pit Bull Politics and Multispecies Justice*. Seattle: University of Washington Press.

Weheliye, Alexander G. 2014. *Habeas Viscus: Racializing Assemblages, Biopolitics, and Black Feminist Theories of the Human*. Durham, NC: Duke University Press.

Winter, Christine J. 2020. "Does Time Colonise Intergenerational Environmental Justice Theory?" *Environmental Politics* 29, no. 2: 278–96.

Winter, Christine J. 2021. *Subjects of Intergenerational Justice: Indigenous Philosophy, the Environment, and Relationships*. London: Routledge.

Wise, Steven. 1997. "Hardly a Revolution—The Eligibility of Nonhuman Animals for Dignity-Rights in a Liberal Democracy." *Vermont Law Review* 22, no. 4: 793.

Wolfe, Cary. 2003. *Animal Rites: American Culture, the Discourse of Species, and Posthumanist Theory.* Chicago: University of Chicago Press.

Wynter, Sylvia. 2006. "On How We Mistook the Map for the Territory and Reimprisoned Ourselves in Our Unbearable Wrongness of Being, of Désêtre: Black Studies toward the Human Project." In *Not Only the Master's Tools: African American Studies in Theory and Practice*, edited by Lewis R. Gordon and Jane A. Gordon, 107–69. New York: Paradigm.

Sophie Chao is Discovery Early Career Researcher Award (DECRA) Fellow and Lecturer in the Discipline of Anthropology at the University of Sydney. Her research investigates the intersections of ecology, Indigeneity, capitalism, health, and justice in the Pacific. She is author of *In the Shadow of the Palms: More-than-Human Becomings in West Papua* (2022) and coeditor with Karin Bolender and Eben Kirksey of *The Promise of Multispecies Justice* (2022). For more information, please visit https://www.morethanhumanworlds.com.

Danielle Celermajer is a professor in the discipline of sociology and social policy, deputy director of the Sydney Environment Institute, and lead of the Multispecies Justice Collective. Her research focuses on justice, responsibility, and imagination and the multidimensional unraveling of a climate-changing world. Her books include *Sins of the Nation and the Ritual of Apology* (2009); *The Prevention of Torture: An Ecological Approach* (2017); *A Cultural History of Law in the Modern Age* (2019), with Richard Sherwin; *The Subject of Human Rights* (2020), with Alex Lefebvre; *Institutional Transformations; Imagination, Embodiment, and Affect* (2021), with Millicent Churcher and Moira Gatens; and *Summertime: Reflections on a Vanishing Future* (2021).

STYGOFAUNAL WORLDS

Subterranean Estrangement and Otherwise Knowing for Multispecies Justice

Astrida Neimanis

Abstract **How can we cultivate an underground multispecies justice with beings whose lifeworlds are unknown and unknowable? This article examines this question through a consideration of stygofauna: miniscule deep-time creatures who make their home in the watery seams of the earth. Taking a cue from these critters—many of whom have evolved without eyes to make their way differently in the darkness of their watery subterranean homes—the article troubles the assumption that knowledge, care, and justice must be predicated on a kind of knowing that insists that humans *literally* bring other worlds to light. Through a specifically situated exploration of stygofaunal worlds, knowledge, and mining in Australia, the article asks, How is knowledge-as-illumination complicit with complex regimes of knowledge where knowing in the name of justice is tangled up in knowing as a further (colonial, speciesist, ableist) violence? Refusing purity politics, the article's first aim is to demonstrate our complicity with extractive knowledge regimes even in a quest to care for underground worlds. Second, the article insists that knowing otherwise is both possible and already at work. It argues that to know stygofauna *otherwise*, one cannot eschew science or knowledge altogether. Instead, it proposes that multispecies justice depends on two moves: first, on safeguarding a mode of unknowability that the article refers to as estrangement, and second, on recognizing and cultivating knowledge practices that can cultivate nonextractive relations with subterranean species, even if imperfectly. It concludes with a short overview of several examples of knowing otherwise that push readers to think differently about knowledge as a practice of care and justice.**

Keywords **multispecies justice, stygofauna, underground, knowledge, extraction, ocularcentrism, Australia, mining, science, environment, groundwater**

Cultural Politics, Volume 19, Issue 1, © 2023 Duke University Press
DOI: 10.1215/17432197-10232445

1. Introduction: Extraction and/as Knowledge Practice

Extraction is a complex phenomenon on which whole worlds are built and others are negated. In Australia, where I lived for half a decade, the issue of extraction is everywhere and nowhere. It is everywhere in that Australia is among the top five producers of most of the world's key mineral commodities. Mining is a pillar of the economy, accounting for 10 percent of the GDP in 2020 and growing (Zakharia 2020). Tellingly, during the 2020 pandemic shutdown, miners were classified as essential workers (and business leaders in extractive industries were called on by the country's prime minister to chair the pandemic economic "gas-led" recovery plan). Even though the mining sector only employs about 2 percent of the total workforce in Australia, we are all heaving under the weight of its spoils: bricks, glass, cars, computers, batteries, lightbulbs, fertilizer, paint, coins, medical equipment, cutlery, bathroom taps, bridges, bikes, blenders, mobile phones, dental fillings, jewelry, aluminum foil, nail clippers, and so on. But for many of us extraction is also nowhere. Living in a large urban center like Sydney, one might believe that all of these items are magically conjured; our everyday use of them seems to have little to do with unsafe levels of lead in a regional town's drinking water, catastrophic drawdown of World Heritage–listed wetlands, desecration of sacred Indigenous sites—or with the possibilities for an underground multispecies justice, which provides the overarching context for this article.

In the face of such unwieldy entanglements, one scholarly approach suggests beginning from where you are, with what you can (see, e.g., Dumit 2014). In the shadow of extraction, however, even this approach gets tricky: the majority of mines are located at a great distance from me; the human and more-than-human lifeworlds that these operations devastate exist in places that are not only imaginatively and geographically remote but also exceptionally diffuse. Like climate change, or patriarchy, or capitalism, or extinction, or militarism, extraction is an amorphous phenomenon that we are acutely aware of yet whose specific coordinates are often difficult to pinpoint. We do not want to look away, but where specifically do we look? What specificity could appropriately signal the palimpsestic effects, the time travels and deep channels that keep this phenomenon at a distance, even as it clothes us, transports us, connects us, feeds us? Indeed, my fingers type these words on a silver Apple wireless keyboard made from anodized aluminum. I learn this via a quick flick over to Google, where I also discover that Australia is the largest producer of bauxite in the world. Eschewing any environmental purity politics (Shotwell 2016), we must acknowledge that our complicity runs deep, even as our knowledge of these implications is rather shallow.

We (a "we" which this article interpellates as its most probable reader: an educated, mostly urbanized and middle-class audience, even as "we" are also differently situated in key ways) could say that this scalar torquing of time, distance, and mattering, whether by accident or intent, is what keeps us looking away. As such, these unwieldy tangles also provide the necessary sustenance for an ongoing capitalist, consumerist, and colonial approach to the underground. But if "looking away" or not seeing is the problem, then what is the solution?

Rather than trying to hold the entire extractivist apparatus in view, here I parse these questions through the lifeworlds of one kind of animal known as stygofauna. Miniscule deep-time creatures who make their home in the watery seams of the rock, stygofauna actively resist certain kinds of knowledge. I ask what thinking with these groundwater-dwelling species under threat from extraction might teach us about not looking away, when *seeing* is not easily accessed and in fact rendered problematic. Taking a cue from these critters—many of whom have evolved without eyes to make their way differently in the darkness of their watery subterranean homes—I trouble the assumption that knowledge, care, and justice must be predicated on a kind of knowing that insists we *literally* bring other worlds to light. To look, after all, is associated with a privileged ocularcentrism where to see is to know (see, e.g., Jay 1988; Celermajer 2006), and to know is to make care and justice possible. Yet despite long-standing critiques, a reliance on seeing persists. For example, as Petra Tschakert (2020: 15) argues, political encounters that lead to more-than-human solidarities as a basis for multispecies justice "also require visual, embodied, and ethical engagements." While justice is about more than seeing, Tschakert contends that this is where it begins. I trouble this assumption through a specifically situated exploration of stygofaunal worlds, knowledge, and mining in Australia. I ask: How is knowledge-as-illumination complicit with complex regimes of knowledge where knowing in the name of justice is tangled up in knowing as a further violence? That is, what are the ethical implications of equating sight with knowledge, particularly in relation to mastery, colonialism, and ableist anthropocentrism as accomplices to a subterranean multispecies injustice?

This article thus grapples with how multispecies justice in the context of extractivist regimes is bound up in both knowing and not knowing. In a continued refusal of purity politics, my first aim is to demonstrate our complicity with extractive knowledge regimes even in a quest to care for underground worlds. Second, I insist that knowing otherwise is both possible and already at work. After an overview of stygofaunal lifeworlds, I briefly survey the burgeoning interest in the underground within environmental humanities and multispecies studies more generally. I then turn to the complex ways stygofaunal worlds come to be known, by whom, and to what end. I examine how these forms of knowledge are differently implicated in the Western and colonial impetus to knowledge as visibility as mastery, but in ways that are hardly straightforward. While "bringing to light" information about these species and their lifeworlds might seem a welcome step in addressing extraction-related multispecies injustice, we discover that some knowledge is complicit both in addressing *and* perpetuating violence, and sometimes both at the same time. The nature of knowledge never becomes fully visible: knowledge surfaces, then burrows again, finding subterranean complicities that refuse any simple separation between good and bad ways of knowing.

Following this, I turn to ways of knowing stygofauna *otherwise*. These do not—nor should they—eschew science or knowledge altogether. Instead, I propose that multispecies justice depends on two moves: first, on safeguarding a mode of unknowability I refer to as estrangement. I draw on what Marisol de la Cadena (2015) has named the "Anthropo-not-seen"—worlds that cannot be fully subsumed by Western knowledge—as well as eco-crip approaches that call for bodily diversity in ways that deprivilege ocularcentrism. The second move is to recognize and cultivate ways of otherwise knowing, by which I mean knowledge practices that can cultivate nonextractive relations with subterranean species, even if imperfectly. I conclude with a short overview of several examples of knowing otherwise—including different Indigenous water management practices and two works of fiction—that push us to think differently about knowledge as a practice of care and justice.

2. *Stygofaunal Worlds*

Stygofauna are tiny animals, largely invertebrates, that live in subterranean waters. Named after the river Styx (that underwater stream that carried Greek mythology's dead souls from one world to another), most stygofauna are "unpigmented, elongate and small, adapted for life in dark and often confined spaces" (Tomlinson et al. 2007: 1317). Or, in the memorable words of Western Australian stygofauna scientist Bill Humphreys (pers. comm., 2015), we might consider them "small white and blind cockroaches that get in the way of mining." They mostly comprise crustaceans but also include beetles, snails, mites, and worms. They have evolved to exist without sunlight, in constant temperatures, and are dependent on infiltration of nutrients from the surface world. Because of this, stygofauna are most abundant in shallow aquifers where food supply and oxygen are generally more plentiful.

Ironically, Australia's aquifers were until recently thought to be a biological desert. Yet the diversity of this continent's stygofaunal populations is in fact among the richest in the world, with known taxa numbering in the thousands. The biodiversity of Australian stygofauna is related to their narrow endemism, which means that particular species are often found only in very localized habitats; they have very limited spatial distributions; and their diverse communities are narrowly bounded. In Australia, the environmental value of stygofauna is compounded by the continent's geological record. Quite simply, these groundwaters are very, very old. Many species would fulfil International Union for the Conservation of Nature (IUCN) criteria for listing as vulnerable or endangered due to their aerial extent alone, or because some species are restricted to a single, localized community. Related to this, their deep dwelling in place also makes them "important subjects in unravelling deep history" (Humphreys 2006: 115). Their groundwater homes have come to be known as "living museums" (115) and their tiny bodies "relicts" (124), offering clues to deep-time worlds. This temporal-spatial idiosyncrasy and isolated genetic diversity bestows on them additional scientific and conservation significance—another (extractive, colonial) version of the "lost tribe," whose DNA are mined in hopes of coaxing secrets out of a planetary deep past.

Stygofauna are moreover responsible for unnoticed labors that maintain the surface world. Eating bacteria, they keep the groundwater clean. They graze biofilms, modify biogeochemical properties, and alter interstitial pore size in aquifers with their limited albeit consequential movements. They physically transport material through the groundwater environment. Aquifer health depends on these small labors, and surface water health depends on the well-being of subterranean waters (Humphreys 2006; Guzik et al. 2010). Like a small crustaceous canary, stygofauna are also considered biological sentinels; they "offer promise as indicators of groundwater quality" (Tomlinson et al. 2007: 1318). This emphasis on the work they do for the benefit of others is not to instrumentalize them but rather to stress that multispecies justice in the context of extraction is always relational. While these creatures must have value in and for themselves, we come to understand them because their lifeworlds intersect with ours, even if at a remove. In this sense, we might consider them "shadow bodies" in the manner suggested by Australian feminist ecophilosopher Val Plumwood's (2008) concept of shadow places: the subterranean waters are the shadow places where the labors of these shadow bodies go mostly unnoticed. However, while in Plumwood's concept of shadow places the prosperity of dominant places depends on toil and injustice within the shadows, stygofaunal worlds model the *possibility* of a more reciprocal relation. While the above- and belowground worlds shadow one another, their welfare is codependent.

Most of the literature on stygofauna stresses their reliance on the even keel of their subterranean climate. They are adapted to a near steady-state environment which means that their well-being is particularly susceptible to changes in water quality, where groundwater deviates from background conditions. Their capacity to recover from such disturbances is limited by low mobility, low reproductive rates, and dependence on localized and limited food supplies (see Humphreys 2006; Guzik et al. 2010; Hose et al. 2015). While some recent studies suggest that the "steady state" required by stygofauna is exaggerated—after all, groundwater conditions are in some degree of flux all the time—it is also clear that rapid and excessive incursions into their habitats can be devastating. Turning our attention back to the surface, we see that the extractive industries that are damaging aboveground worlds are also violently puncturing these subterranean ones, to equal devastation.

In other words, disturbances aboveground will be shadowed by disturbances belowground. Massive mineral extraction industries in Australia, and mining and coal seam gas developments in particular, pose a threat to stygofaunal well-being (Hose et al. 2015), which qualitatively and quantitatively affects aquifer health. Groundwater extraction including mine dewatering significantly alters the sediment matrix necessary for stygofaunal nourishment. Subjected to groundwater drawdown caused by extraction, stygofauna can become stranded, or forced to penetrate the earth more deeply—where nourishment is in short supply. They are also sensitive to changes in aquifer pressure, and pore dimensions due to subsidence. Moreover, the effects of coal seam gas developments leak far beyond the target aquifers to impact contiguous groundwater. Contestations related to extraction in Australia encompass multiple concerns, from contaminated rural drinking water supplies to the destructions of sacred Indigenous sites, to farmlands ruined by toxic seepage. In this context, stygofauna are just one of the more diminutive and lesser-known victims of extractivism in Australia.

Under the shadow of modern extraction industries, stygofauna are marooned, starved, and displaced in a polluted and drying underworld. Attention to stygofauna, groundwater and extraction thus reminds us that anthropogenic climate change as spurred on by fossil-fueled extractive industries is also occurring below the surface of the earth. We begin to feel the rumble of the weather underground: temperatures changing, hydrogeomorphologies shifting, populations migrating or being incapable of doing so. Climate change is not something we know only through carbon emissions or rising temperatures experienced by humans and directly affecting our lifeworlds but something we also come to know by paying attention to the worlds of nonhuman species, including stygofauna. Yet given that their shadow dwellings are almost entirely inaccessible to us, *how* we come to know this is not an innocent question. The next section considers the unknown aspects of stygofauna and the groundwater habitats, and the desire to transform this uncertainty into *more* knowledge by "bringing things to light."

3. Surfacing Knowledge, Bringing the Underground to Light

There is much about stygofauna that we don't know. Despite the recent increase of information about their ecological significance, the scientific literature uniformly acknowledges that this data barely scratches the surface (see Humphreys 2006; Thoo 2012; Mokany et al. 2018). There is no natural history of stygofauna in Australia, for these species do not readily give themselves over to observation, notation, or contact. On some estimates, maybe 10 percent of their taxa in Australia are known to us. How far do their communities extend? How deep? These answers remain mostly guesswork.

This uncertainty is due in part to the "unique constraints" (Korbel et al. 2017) of groundwater sampling and the physical inaccessibility of subterranean worlds (to us). It is moreover compounded by the "subsurface heterogeneity" of groundwater ecologies and the speciation of stygofauna in these contexts. Stygofauna species, we recall, have very limited range. Thriving species that you discover below your feet may be complete no-shows just over the next hill. Their narrow endemism means that extrapolations and generalizations are not much more than a poke in the dark. Definitive knowledge about groundwater places in which stygofauna dwell (aquifers, karst systems, layers of shale) is similarly elusive. Groundwater cannot be measured with certainty, as knowing groundwater is not a simple matter of collating different numerical values, such as time, space, volume, consistency, depth. For example, although we can measure groundwater flow through an application of Darcy's Law (a mathematical formula used to by hydrogeologists to describe the rate of flow of water through a substrate),[1] each of its constitutive elements is variable, increasing exponentially the complexity of any one calculation. In other words, not only is there wild diversity in terms of where and how stygofauna inhabit the underground, those underground homes themselves are wildly uneven and difficult to measure.

This difficulty is met, unsurprisingly, by calls for more knowledge, more thorough knowledge, and better knowledge. Noting that "successful conservation and management of groundwaters and their dependent ecosystems rely on better public understanding" of stygofauna and their ecologies, leading Australian stygofauna researchers aver that "further research on subsurface processes and taxonomy, and legislative protection of rare and threatened subterranean communities and species" are necessary (Boulton et al. 2003).

Putting our ear to the ground, we hear echoes about this lack of awareness in terms of subterranean worlds more generally. This includes emerging scholarship in cultural geographies and environmental humanities. In a pivotal text titled "Unearthing the Subterranean Anthropocene," Marilu Melo Zurita, Paul Munro, and Donna Houston (2017) call attention to the discursive absence of the underground from contemporary debates about the

Anthropocene. The authors seek to challenge environmental humanities and social sciences' "surface bias" and highlight the ways in which our (primarily) aboveground lives are entangled with the rock stratum. The authors note the importance of bringing these relations out of obscurity (1). Calling for more ethical relations to underground worlds, they aver that "the need to *unearth* the Subterranean Anthropocene and its more-than-human temporalities is vastly important to the futures of multitudes of earthly inhabitants" (5; emphasis added).

Here, we see a scientific desire for bringing stygofauna to light coalesce with a social scientific call for (metaphorically) illuminating subterranean worlds as meaningful spaces more generally. In both cases, the "unknown" quality of the underground is framed negatively, and the invisibility of these worlds is cast predominantly as a problem (to be solved, to be overcome). This corresponds with a long-standing association in Western philosophy between knowledge and illumination, as explained by political theorist Danielle Celermajer (2006) via her parsing of the philosophies of Hans Jonas and Hannah Arendt. This is also the well-known argument advanced by feminist science and technology studies scholar Donna Haraway (1988) in her critique of the "view from nowhere." In these critical appraisals, vision is directly connected to mastery, as "the subject is the master of the object he chooses to see" (Celermajer 2006: 134). To see is to know, and to know, in this version of knowledge, is to master. This argument dovetails with what Martin Jay (1988: 306) calls a "deep-seated distrust of the privileging of sight" within Western thought. As Celermajer details, seeing as knowledge also connects with the treatment of others as objects, where the seeing subject will treat the world as separate from the unimplicated self. Knowing in this way means that we can portend objectivity (as we remain separate from what we know). Or as Haraway (1988: 581) puts it, this is the kind of vision "honed to perfection in the history of science tied to militarism, capitalism, colonialism, and male supremacy—to distance the knowing subject from everybody and everything in the interests of unfettered power."[2] Rather than objectivity, then, this is an exercise in *power over*. In reading Jonas's work, Celermajer also corroborates this: "the seeing subject . . . might choose to rest her eyes where she wishes, or even to use her eyelids to block out objects altogether" (135). Thus this master knowledge is also the rejection of ethical relation in which the "seen" can be known on its own terms.

Turning back to stygofauna and the underground then, it is curious that despite long-standing critiques of ocularcentrism, our impetus is again to unearth and bring to light. On the one hand, we can understand this as an ethical gesture—we are choosing "to rest our eyes" on the underground as a way to counter our ignorance. But this move cannot bypass vision as also that form of knowledge that resists relationality and imparts mastery. Here, we have to remember that the desire for metaphoric illumination of stygofaunal lifeworlds merges with a literal impetus to unearth these stygofauna—pulling these animals up and out of their groundwater darkness. Moreover, even this physical incursion into groundwater environments does not lead to certainty. For example, hydrogeologist Emma White (2018) notes that by drilling "at only a few spots in the study area," these locations "are then used to extrapolate what is underground across the whole area—a bit like a dot-to-dot puzzle. But sometimes it can be like trying to sketch a whole person when all you can see is their little toe. That is to say, we just don't know for sure." White cites the example of the Adani Carmichael coal mine project. During the approval process, Adani's own modeling of the mine's effects on Doongmabulla Springs—a groundwater system with high ecological and cultural significance—relied on a "best-case scenario" analysis. White points out that in relation to groundwater environments where so much is unknown, an "uncertainty analysis" is vital for good decision-making: this analysis considers all plausible scenarios and not just the "best-case" ones. Nonetheless, respect for the underground's unknowability is still overshadowed by calls for *more illumination*: a lack of "critical scientific data" contributes to "uncertainty and conflict"; as a result, the scientists who contested Adani's modeling call for a "higher standard of scientific information to be collected and reviewed." In the specific case of the Adani Carmichael coal mine, this would have meant deferring any approvals "until the data gaps responsible for the uncertainty were filled" (Currell et al. 2017). Again, we are back to the desire for research to bring more knowledge (and more bodies) to light on our own terms.

What we need is to bring philosophical critiques of ocularcentrism into relation with scientific demands for more knowledge—and knowledge of a particular kind. In turn, these need to be contextualized within anticolonial engagements with both: in settler contexts, "settler-ocularcentrism" (Wesner 2018: 2) can still shore up colonial investments in "nature," just as science "in the name of" conservation can still shore up settler colonialism's persistent assumed access to Land (Liboiron 2021). With these considerations in mind, although we might interpret the call for knowledge-as-illumination as a way to check extraction industries and curtail their incursion into delicate ecosystems, we must also be alert to the violences in which these calls are complicit. For instance, CSIRO (Australia's Commonwealth Scientific and Industrial Research Organisation) points out that "the dearth of knowledge surrounding the diversity, distribution and ecology of stygofauna in Australia creates considerable uncertainty in the assessment of ecological risks associated with coal mining and CSG activities" (Hose et al. 2015). In an era of high anthropogenic extinction rates, biodiversity estimates can identify knowledge gaps for the purpose of developing conservation policies. But on the other hand, other kinds of intimacies are also at work; the CSIRO report authors write that "research and new knowledge about stygofauna will inform and improve the risk assessments for groundwater ecosystems that are increasingly necessary *as part of* mining developments." This research "will *promote the coal industry as a leader in the sustainable management* of aquifers and biodiversity" (vi; emphases added). In one quick slip down the drill hole, mining corporations are now also the guardians of subterranean ecological well-being.

This slippage aligns more generally with extractive capitalism's desire to illuminate the underground. For example, coal seam gas corporations in northern New South Wales (Australia) hope that "the invisibility of underground water flows would be overcome at the same time as another carbon producing fuel is accessed" (McLean et al. 2018: 625). Such examples pepper yearly reports and websites of extraction corporations. For example, the pop-up message on the *Australian Mining* magazine website flashes, "Don't be in the dark! Subscribe to our magazine now!" In other words, the underground is (still) a new frontier. The overlay of metaphor and materiality in bringing the underground to light illuminates something else too—namely, a close proximity of Western knowledge projects (within both natural and social sciences) keen to cultivate closer relations with underground worlds, with colonial capitalist undertakings.[3]

Further intimacies are revealed when we loosely map extant knowledge about stygofauna in Australia onto major extractive projects here. As I dig through stygofauna conservation reports, I observe something quite curious: areas of greatest stygofaunal diversity seem to coincide with major resource extraction activities. While I am first alarmed ("No! Don't drill, and especially don't drill *there*!"), I realize that it is not just diversity but Western scientific knowledge about stygofauna in general that maps quite directly onto a geography of known coal and gas seams as well as uranium deposits. The logic becomes obvious: the former is dependent on the latter. As scientific research on stygofaunal diversity notes, "Regulations requiring the inclusion of subterranean fauna during the environmental review process for major resource prospects . . . by the Environmental Protection Agency have accelerated the discovery of new species" (Guzik et al. 2010: 409). Similarly, research into effective sampling techniques also concludes that "the vast majority of data for subterranean fauna in the Pilbara [an area of rich stygofaunal diversity] have been collected as part of biological surveys undertaken for environmental impact assessments of proposed mining developments" (Mokany et al. 2018: 241).

To state this more plainly: without extractivism, we would know almost nothing of these underground worlds. It is large-scale extraction that allows us to extend our attention to these lifeworlds below. It is only *because* of assessments undertaken to facilitate drilling down that we know much of anything about stygofauna at all. Or, to nuance this claim in a key way: without extraction, we would have almost no Western scientifically based evidence of stygofaunal lifeworlds.

Where does this leave us? In the first place, thoroughly complicit with settler-colonial knowledge-making in service of extraction. Seeking multispecies justice can sometimes be framed as a quest for intimacy: a desire to know other bodies—to "rest our eyes" on them in the hopes that this knowledge will make us care more. But while intimacy does not necessarily mean mastery (Neimanis 2017: 145–46), the dubious power of "seeing" has already been questioned in the name of seeking justice for nonhuman animals (Rasmussen 2015; Giraud 2021) and is moreover closely aligned with a colonial, extractivist project whose intimacy with Western and colonial scientific knowledge projects is revealed in their shared desire to "bring to light." *These overlaps need to be taken seriously.* What still needs to be thought, when the quest to do justice to more-than-human lifeworlds suffering under and resisting extractivist regimes is always bound up in extractivism's own devastations? My suggestion is that this kind of doing justice also risks slipping into mastery and potentially denying the possibility of ethical relation—not because we aren't looking at these other beings but because precisely in insisting on looking at them, we might destroy their lifeworld possibilities. In the final section, I will come back to suggest other kinds of knowing as relation that can resist this kind of seeing. In the next section, though, I burrow more deeply into the overlap between knowledge and extraction, following their subterranean channels of connection and complication.

4. Anthropocenic Knowledges as Extractive and Never Enough

Importantly, the underground—including stygofaunal lifeworlds—has not been always and everywhere ignored or imagined as empty. Anthropologist Andrea Ballastero (2019) refers to the deep and vivid imaginaries that undergrounds elicit through time; these subterranean places are described as mysterious, with mines that eat people and caves full of spirits and promises. Melo Zurita (2020) describes Mexico's cenotes in similar terms, where these underground waterscapes have been crucial elements of human cosmologies for millennia—they are places of "respect and veneration," "a place from which life emerges but that takes life as well." Heidi V. Scott (2008: 1854) points out that even in the context of colonial expansion (which, she notes, has always had its sights set on the underground), "the ways in which Europeans imagined, used and attributed meaning to the colonial underground were far from unitary or homogonous."

Such examples underscore the myriad ways in which undergrounds have been made meaningful in ways not necessarily related to extraction. In other words, subterranean knowledge itself is not the problem. The problem is rather the kind of knowledge that these projects support—one that I refer to here as a "modern" kind of knowing that aligns with an "Anthropocene" imaginary of the underground (see Linton 2012).[4] Such an imaginary renders nonhuman elements knowable through a particular kind of order. Like water that via the hydrological sciences and their colonial capitalist context has come to be seen as atomized, exchangeable, and fungible in order to become "Anthropocene water" (see Neimanis 2017, chap. 4; Linton 2012), or like rock that via geology's classifications comes to be treated as commodifiable property (see Yusoff 2018), subterranean lifeworlds too can be "Anthropocenized." Here, the underground is emptied out of meaning in order to be filled back up again by managed, measurable, and ultimately "master-able" matter. For example, in a more recent article on "sub terra nullius," Melo Zurita (2020: 270) again takes up the "discursive subterranean nothingness" that she describes as a "physical underground space . . . being epistemologically rendered as a blank space." She argues that like the "terra nullius" imaginary that occupied the Australian continent with European invasion, today's underground is also by necessity imagined as empty in order to be able to bend it to the will of capitalist colonialism. Empty, of course, does not mean empty of literal things—the point is rather that it is empty of meaning for the Anthropocene subject. In Melo Zurita's argument, the "knownness" of the underground only emerges if and when it can be made intelligible in terms of geological, hydrological, and technological measure.[5]

Melo Zurita, Munro, and Houston (2017: 4), drawing on the work of Bruce Braun, likewise recount how the making legible of the underground was directly linked to projects of economic, social, and political calculation. Anthony Bebbington (2012: 1160) similarly notes that our current capitalist formations were "made possible by the bundling of the underground with specific networks of power, knowledge, and technology." Scott (2008) also underscores the particular role that mining has played in the establishment of the "colonial underground," while Melo Zurita, Munro, and Houston (2017) argue that urban development is increasingly part of this vertical expansion. Building on these arguments and following Melo Zurita (2020: 271), who is concerned with the "social, political and financial effects of its perceived emptiness," whereby "value can be extracted from claims of nothingness," I am similarly interested in how, via extractive knowledges, all of the unknownness of stygofaunal worlds is filled in ways that complicate any project of multispecies justice. Here, we might look more specifically at the information gap around stygofauna noted in the previous section. We need *more* information to bring these imperiled critters into the light, both social and natural scientists plead. But such calls to "unearth" the Anthropocene or to surface stygofaunal worlds raise the question: What are the stakes of "knowing more" about stygofauna and their groundwater worlds?

As intimated above, those interested in environmental well-being and multispecies justice could take these pleas at face value and agree that an evidence base from which to make conservation and ecological management decisions is just *good practice.* Following the deficit model of science communication, knowing more gives us access to better grounded arguments; we can use this information to challenge the claim that there are no significant environmental consequences to extraction, or none that cannot be readily remediated. That is: in a time of ecological crisis, we need to *surface the facts*—where those facts adhere to Western scientific principles. This is indeed the recommendation of hydrogeologists and other scientists interested in the ecological well-being of the underground. For example, in their study on the inefficacy of EPA surveys in accurately describing the threat to stygofauna from mining activities, Tomislave Karanovic and coauthors recommend "that in geographic regions and geology and aquifer types where stygofauna are not diverse or abundant, repeated sampling of many bores is required to confirm the presence of stygofauna" (Karanovic et al. 2012: 564). More samples equals more knowledge.

But the knowledge tunnels get murkier, again, when we examine how this "more knowledge" is generated, and by whom. While it may seem odd that mining corporations themselves would show any interest in stygofaunal well-being, the Groundwater Dependent Ecosystems Management Plan for the proposed Adani Carmichael coal mine, for example, which was given the green light by the government in 2019, mentions stygofauna a whole thirty-one times! Monitoring "stygofauna presence and endemicity," it reads, will be a crucial part of ascertaining that the corporation has indeed "minimized the impact of aquifer drawdown." Their means proposed for doing so? "Stygofauna survey."

What are we to make of this? The first point has already been discussed: more knowledge means literally more incursion into stygofaunal lifeworlds, as sampling and surveying requires drilling holes into stygofaunal habitats—worlds that are explicitly harmed when their steady state is disturbed, and worlds that are so diverse, existing mostly in very local ways, making extrapolation difficult. The second point, though, is about what this promise of "more knowledge"—that Adani pledges and that scientists interested in conservation also call for—actually illuminates. If we look more closely, we see that this purported generation of knowledge in fact submerges further meaningful understanding. To paraphrase the Adani report: *We are going to ensure the well-being of stygofaunal communities through stygofaunal surveys.*[6] This move first serves as an alibi for the extractive industry (for it is on the basis of this promise—and not on what information any such survey might actually yield—that drilling is allowed). Second, as noted, such surveys also harm the very communities they seek to ostensibly protect. "Monitoring" well-being by making increasingly more incursions into the underground is questionable. But, third, we might also ask: Why are such surveys needed? All of the scientific evidence already points to the fact that these communities are ecologically valuable, genetically precious, and also vulnerable in the shadow of climate change and extraction. What more would additional surveys reveal? The production of knowledge here does not perform its stated intention—namely, care and concern for these communities.

We could look at the case of the Yeelirrie Uranium mine in the western Australian outback. As noted by environmental engineer Gavin Mudd (2019), stygofaunal biodiversity here is exceptionally rich and even prompted a 2016 EPA ruling that deemed "there was too great a chance of a loss of species that are restricted to the impact area." The Tijwarl traditional owners have been fighting the mine for over forty years. In spite of this, the Cameco mine was approved in 2017. Noting the potential environmental impact, the then minister of environment of Western Australia, Albert Jacob, assured those concerned that the situation would be closely monitored: "Further surveys may identify that the species currently only found within the project area are more widespread. I have therefore mandated as part of this approval further survey work and investment in research" (*ABC News* 2017). Even though ample knowledge attested to the imminent extinction of numerous stygofaunal species, the decision was again upheld in 2019 by the supreme court. In other words, knowledge here becomes increasingly removed from its stated intention: incursion after incursion, and extraction after extraction, and yet the information cannot perform its stated function, to preserve biodiversity. In such circumstances, we must again ask: What is at stake when we call for "more knowledge" in the name of multispecies justice?

There are ways to attempt gathering all of the information we might need to "know" stygofauna and their groundwater homes, but each of these steps demands a progressive incursion, a deliberate bringing-into-visibility that also violates that which we seek to know. In the words of Australian groundwater scientists, "It is futile to wait until all biodiversity and threatening processes have been identified before acting to protect GDEs [groundwater dependent ecosystems]"; this would mean advocating "protection aimed at the habitat or ecosystem scale" (Boulton and Humphreys 2005: 51). There is here both a tacit acknowledgement that full knowledge cannot be the prerequisite for ethical action and also an understanding that a world contains many worlds—a point I return to in the next section.

Gathering data in order to protect these vulnerable species becomes a never-ending quest toward never-enough: no amount of knowledge about stygofauna seems to be enough to stop mineral and fuel extraction in their habitats, unless, of course, we were to literally sample every square meter of ground, unearthing every hidden corner. (Imagine: all those punctures . . .)[7] Is this fossil-fueled complicity the only thread—traced by the fractured, stony splitting of the mineral seam—that allows us to get to know stygofauna? If "knowing each other" is our ground for multispecies justice, what does it mean that extraction—as industrial energy project but also as knowledge project—provides the conditions of this possibility? The mining corporation, that tiny translucent body without a backbone, and us: cleaved together through the wound of the earth, where cleaving is both an attractive pulling together and a violent pulling apart.

5. Otherwise Knowing: The Anthropo-not-seen, Not Seeing, and Estrangement as a Tactic for Multispecies Justice

In an Anthropocenic imaginary, then, knowledge of the underground is measurable and knowable, determined by an episteme of illumination that holds what is known in objective, masterful separation. As we see above, this relationship to knowledge is promoted by scientists and exploited by extractive industries. Both those Anthropocenic knowers who want to protect stygofauna *and* those who wish to tear up the earth seek more knowledge, but in neither case is this a guarantee of more just multispecies relations with underground worlds. In fact, in its inability to relinquish mastery, this can even exacerbate material violences.

This brings to mind what anthropologist Marisol de la Cadena (2015) calls a translation of entities into a "universal nature." Anthropocenic knowledge projects are extractive, and result in the levelling of worlds. This is a metaphorical leveling, as through this razing, worlds are subjected to a hegemonic value system and a single currency of knowledge. This aligns with Anthropocenic knowledge as measurable, reducing matter to its fungibility but also reinforcing its "never enough-ness." But the leveling is also literal, in terms of the "unprecedentedly unstoppable—and mighty—corporate removal of resources in places formerly marginal to capital investment" (de la Cadena 2015).

On de la Cadena's understanding, it is not possible to persist entirely *outside of* these conditions: we (in the full multispecies interpellation of this pronoun) are *both* all in this together *and* we are all in this together in very different ways (Braidotti 2019; Neimanis 2017). However, de la Cadena also insists on an "Anthropo-not-seen" where some modes of existing are *in excess* of this leveling. This underbelly of the Anthropocene (sharing its territory, sharing its nomenclature) signals the universalizing error of this Age of Man but also insists that there is an excess that cannot be extinguished, assimilated—or fully mastered. Some things cannot be properly known not only because "we" humans cannot easily penetrate the uneven, miniscule, filigree crevices of the watery earth with our own bodies, nor only because do not have the right technoscientific means (yet) to do so, but because some worlds do not easily give themselves over to dominant Western ways of knowing. De la Cadena's crucial point is that the Anthropo-not-seen is "the destruction of worlds *and* resistance to it." The "not-knowing" of stygofaunal worlds is a resistance to certain kinds of knowledge: by human eyes, by human hands, by human excavators, by human audits, by human technoscientific measures.

So what are we to do when, as Tschakert notes, multispecies solidarity in the name of justice requires that those nonhumans that are distant to us be brought into some kind of proximity? It seems that in the shadow of extraction, we need *both* to know stygofaunal worlds *and* to honor or match *their* resistance with a resistance of our own: a resistance to mastering them through extractive knowledge. Following projects of thinkers such as cultural theorist Julietta Singh (2018), where the imbrication of mastery as "bad" (e.g., colonialism) and mastery as "good" (e.g., linguistic "mastery") is revealed, can we seek ways of knowing and being in relation that are not premised on mastery but rather its very impossibility?

Recall that the literal act of seeing is implicated in this mastery. Vision remains the dominant metaphor for bringing subterranean knowledges to light, just as sight serves more generally as a privileged form of knowledge. Tschakert also manifests this in her calls for fostering more-than-human solidarities in order to promote multispecies justice. Drawing on Emmanuel Levinas, she refers to the "gazing subject" who sees the face of the Other, and reinforces this power of ocularcentrism in the heuristic she presents as a "visual tool" to explain her proposal (Tschakert 2020: 4).[8] Her call is for a political encounter where resistance and contestations can "be made visible" (12) and even "unseen faces" are conjured as ethical subjects who, even if not visible yet, will one day look back at us (13). Tschakert also refers to the use of a multispecies justice lens as a way "to rectify some of the selective blindness that has plagued climate justice scholarship and movements" (3).

Allowing for a degree of metaphoric intent in Tschakert's discourse (a metaphor from which my own work is hardly immune), we could also attend to her ocularcentrism as a way into thinking *more* about what other kinds of knowledge could appropriately "engage with Distant and Unknown Others" as a "prerequisite for prefiguring alternative, non-colonizing, and just pathways into uncertain futures" (3). I agree with her that "it is precisely this commitment to the most distant and different entities in our relational webs for which the immediacy of the engagement is missing that sets multispecies justice and more-than-human solidarities apart from apolitical or post-political climate justice claims and movements" (12). Staying with stygofauna, this would require forging other kinds of intimacies with these subterranean lifeworlds that can still keep some things strange. These could be intimacies that know stygofauna on *their* terms, not ours, and that, in the words of multispecies anthropologist Natasha Myers (2017: 2), help generate "robust forms of knowing (and not-knowing) that can stand both alongside and athwart science." Such intimacies would challenge "the forms of knowing that institute hierarches among life forms, reaffirm Western conceptions of the separateness of humans from nature, continue to mechanise and commodify living processes, propagate militarised narratives and technologies, and secure the continued silencing of [I]ndigenous and local knowledges" (3)—even when these regimes purport to be in the name of justice.

In her book *The Extractive Zone*, feminist decolonial theorist Macarena Gómez-Barris introduces something she calls the "fish-eye episteme." This is both a literal and conceptual view from below the waterline, where the common (colonial, masculinist, terrestrial, surveying, acquiring, opening, splaying, mastering) perspective is disturbed, overturned, and interrogated. In her words, the fish-eye view "looks back upon the extractive gaze from the muddied depths" (Gómez-Barris 2015: 30)—for example, of a camera lens placed in the water. While this strategy of inversion might appear useful in relation to knowing stygofaunal worlds otherwise, we recall that most stygofauna don't have eyes. There is no fish-eye—*no eye!*—here. Despite its decentering of humanism, the fish-eye still remains tethered to a regime of sight. Stygofauna, however, know their world—including its temperature, chemical composition, vibration, and other bodies—through other kinds of sense organs. Stygofauna bodies are in fact perfectly adapted to their environment. The disruption of their environment is the disabling intrusion; the problem is not any problem of their bodies.[9] This material detail about stygofauna can give us a clue of how to resist knowledge through mastery and instead produce knowledge guided by the worlds of those we seek to know. Here, we invite concerns around ableism to congregate with concerns we have already gathered about settler-ocularcentrism and science's will to see/know. Following the arguments of eco-crip scholars such as Sarah Bell and Eli Clare, "encouraging a commitment to biodiversity in nature" must also be a commitment to "embodied diversity" (Bell 2019: 307; see also Clare 2019). Such eco-crip justice approaches remind multispecies justice thinkers and practitioners that a desire to know the underground—even in the name of justice—cannot be about "overcoming" the impairment of lack of sight but rather appreciating ways of knowing otherwise. Vision is not only a metaphor but one of many material relations to the world.

However, learning from stygofaunal materialities about practical solidarities does *not* mean that we should (or could) know the underground exactly the way stygofauna do. Here, appropriation of worldview would become another form of mastery. Instead, the partial unknowability of Anthropo-not-seen worlds, and (multispecies, eco-crip) justice for nonseeing bodies like stygofauna, should be understood as conjoined ambits. Together, they lead us to a tactic for knowing otherwise that I call estrangement. Estrangement is commonly understood negatively—when once-familiar bodies drift apart or fall out. My proposal is to revalue estrangement—the practice of making or keeping strange—as a nonextractive knowledge practice. Estrangement, in this sense, is not the same as nonrelation, and certainly not equal to neglect or disinterest. Again, following Haraway's advocacy for situated knowledge (1988), estrangement accounts for our incapacity for proximity, which is a more ethical relation than the assumption of separation and objectivity. Estrangement recognizes the violence of extractive knowledge and embraces the impossibility of seeing fully or clearly—or at all. While much multispecies theory and philosophy helpfully emphasizes our commonality as a way of coming to care for and seek justice for nonhuman species, here estrangement is a tactic for knowing at a distance, of precisely the kind that Tschakert calls for.

This estrangement has three vital components: first, following from de la Cadena, it acknowledges what we do not and cannot—and perhaps *should not*—know. It desists from mastery. Second, estrangement would cultivate nonextractive forms of knowing that still acknowledge the presence, significance, and inherent value of other worlds, such as stygofaunal ones. Such forms of knowing do not (and should not) reject science but would not *depend* on the settler-ocularcentric, never-sufficient evidence on which Western scientific positivism hinges. Third, estrangement is also always situated: what is strange for one body is an intimate and proximal knowledge for another. In other words, stygofaunal worlds will not be strange to the groundwaters, the rocky habitats, nor to microorganisms and other subterranean flora and fauna with which they interact. They rather remain strange *to us*, the Anthropocenic know-it-alls. Situated knowledge here also acknowledges the necessity of nonproximity as an ethical stance. None of these components would preempt protection of groundwater homes nor promotion of multispecies justice—indeed, they could even enhance these efforts.

I also agree with Tschakert that moving from abstract idea to *practice* is no easy task (3). Fortunately, we need not fantasize about what estrangement via alternative knowledge practices might look like on the Australian continent. Such forms of subterranean knowledges have long been forged by humans at a distance and can sit athwart, but not necessarily in opposition to, scientific practices.

For example, Waanyi novelist Alexis Wright opens her groundbreaking 2006 novel *Carpentaria* (a story partially about the colonial violence of extraction) by describing the ancestral serpent, scouring into the "slippery underground of the mudflats" of what is now the Gulf country of Australia's Northern Territory. Crawling inland and then back out to sea, over and over again, the serpent left in its wake a system of rivers and "a vast network of limestone aquifers" (Wright 2006: 2). Wright conjures the incomprehensible breath of the serpent, rhyming with the tides. As she explains, "The inside knowledge about this river and coastal region is the Aboriginal Law handed down through the ages since time began. Otherwise how would one know where to look for the hidden underwater courses . . . ?" (3).

Acknowledging the complexity of the knowledge practices and Law that Wright describes, I use this example to underscore the power of nonocularcentric knowledges for guiding ethics and politics about how to live with and learn from groundwater worlds. As Wright also describes, this Law teaches those to whom it has been handed down about climatic conditions, fishing, and seasonal changes. She affirms that "it takes a particular kind of knowledge to go with the river, whatever its mood" (3). Knowing anything about this river—a river "that spurns human endeavour in one dramatic gesture" and one whose surface water and groundwater are part of the same body—requires "the patience of one who can spend days doing nothing." This "inside knowledge," where there is "no difference between you and the movement of the water," is an intimacy, but not one readily available to all. Although this Law and these worlds are strange to me (and arguably should remain so), through Wright's novel I can know, at a distance, the power and importance of underwater worlds and their significance in drawing together human and more-than-human livelihoods. This knowledge demands restraint: a refusal to need to prove, and trust in deep-time knowledge that belongs to other Anthropo-not-seens from which I am estranged.

The value of this kind of knowledge is corroborated in some rare instances of contemporary groundwater management in Australia. In their article "Shadow Waters," which (ironically, in the context of my argument) argues for "making Australian water cultures visible," Jess McLean and coauthors (2018: 625) describe Wiradjuri Indigenous knowledges in northern New South Wales "that acknowledge the interactions of surface and subterranean waters, while often simultaneously looking at future impacts of ignoring these relations in the short term." Even in a context where "scientific knowledge of aquifers and recharge rates are far from certain" (625), Lonsdale (a Wiradjuri woman) and other Wiradjuri women cited in the article do not need Western scientific corroboration to understand the importance of groundwater ecosystems. Their sacred knowledge has taught them about the interconnections of ground and surface water, and how it is connected to knowledge in nuanced ways ("when the rivers get sick we get sick people and [this affects] their ability to pass on knowledge"). Across the country, similar kinds of knowledge are called on to care for the Martuwarra (Fitzroy River). As explained by Anne Poelina (professor and Kimberley Nyikina Warrwa Indigenous woman) and coauthors, although mining and fossil fuel extraction threaten water security in the region, Traditional Owners rely on Customary First Law to guide responsible management. Importantly, "although there are knowledge gaps about the hydrology, connectivity between surface water and groundwater *is* certain" (Poelina, Taylor, and Pedrisat 2019: 5; emphasis added); relations with these systems from time immemorial provide otherwise knowing that does not require conclusive evidence to care for these waters and their associated life-forms. Again, I can know about these systems from a distance and trust this knowledge that has its own rigorous methodologies; my estrangement need not be overcome via mastery.

While the above examples rely on different Aboriginal knowledges in Australia, settler knowledges can also practice respectful estrangement and develop nonextractive intimacies. For example, Melbourne-based ecocritic Deborah Wardle (2020: 1) suggests the use of poetics as a path to knowledge of groundwater worlds. Wardle crafts *Why We Cry*, a climate fiction novel, as her own response to the problem of what Tschakert calls knowing "Distant" groundwater "Others." The characters in Wardle's realist fiction come to know their local groundwater through various nonextractive knowledges: childhood memories, stories, embodied encounters, sounds, Indigenous Law, conversation, and various kinds of science. These partial pathways coalesce into something that is enough to warrant protection of these fragile ecosystems. Wardle also offers a second level of knowing—namely, to readers who come to know via the novel's form and language. As Wardle explains, the readers are brought into a relationship with groundwater through affective channels. They learn groundwater "osmotically," in Wardle's words (1), in ways that interestingly eschew ocularcentrism. Wardle describes darkness, touch, texture, sound and multispecies presence all in ways that distinctly background the need to bring these worlds "to light."[10] The strangeness of these worlds and our estrangement from them are held by Wardle's poetics as a vehicle, rather than impediment, for cultivating new kinds of ethical relations.

This is just a small handful of examples of otherwise knowing of groundwater worlds in Australia. For all of their deprivileging of Western scientific empiricism, they still bring us meaningfully closer to Tschakert's Distant Others and enable ethical intimacies. Their refusal of extractive mastery and embrace of unknowability is also distinctly political in the sense that Tschakert calls for: in their various methodologies, they all hold open space for contestation and dialogue. This also resonates with Celermajer's conclusion that knowledge as seeing must be complemented by knowledge as hearing, which limits the degree of separation that the knower can master; it also resonates with Haraway's demand that vision must be reclaimed for feminist and decolonial purposes, too, and that this can happen precisely by insisting on its always deeply embodied and situated nature and being accountable for the power moves that any act of seeing involves. Such knowledges demonstrate that no one kind of knowledge should readily trump all the others; they insist that not everything can be measured. Knowledges in service of multispecies justice can also keep some things strange.

6. Conclusion: Complicit and Estranged, toward an Ethical Relation

Stygofauna may seem too insignificant to matter, even to those of us concerned with multispecies justice, but the ambit of this article is to make them matter more, in terms of what they can teach us about bodies that know and bodies that can be known. In doing so, we have discovered that there is no straight, illuminated path of knowledge through this underground. Instead we find ourselves inextricable from systems of capitalism, colonialism, science, and other lineaments of power that shape extraction in Australia—relations that pull us together and cleave us apart. This complicity, however, is not a signal to master more but rather an invitation to keep some things strange, as a matter of justice.

Moreover, paying attention to stygofauna reminds us that the worlds we see and sense are dependent on all of the "Anthropo-not-seens"—all the worlds that may not be part of our everyday lives but that, like in the mineralized infrastructures of the subterranean stygofaunal worlds, build complex buttresses, supports, and nourishment exchanges for all the worlds that we do see and more readily sense. We are already in relation to these worlds materially, but the goal of this article is to understand these relations in terms of knowledge politics and justice. Perhaps it is fitting that coming closer to the wicked problem of extraction in Australia becomes a lesson in learning restraint. Perhaps in learning how to know less, and differently, also lies a lesson for how to relate to our earth others more generally.

Explicitly welcoming an otherwise knowing of stygofaunal worlds will not bring these worlds (fully) into the light; this is not what they seek. Instead, these ways of knowing acknowledge that other worlds remain just beyond our field of vision, in a desistance from mastery. This deliberate estrangement, a *letting be strange*, is a kind of ethico-epistemology for multispecies justice. It is neither a rejection of nor incompatible with a rigorously scientific point of view.

Instead, knowing otherwise is multivalent, multiversal. It connects to decolonial and poetic ways of knowing that do not rely on scientistic positivism but also to eco-crip justice in acknowledgment of embodied diversity. Attending to these otherwise knowledges, we can trace a fissure of connection between worlds—a relationship that travels along that interstitial highway of the watery rock seam, to create possible solidarity between species, bodies, and worlds, both proximate and distant.

Acknowledgments

Many people contributed to the development of this article. Perdita (Perdy) Phillips and William (Bill) Humphreys introduced me to the world of stygofauna in 2015. Perdy Phillips coorganized a "Going Underground" "trainshop" with me in 2017 to the State Mine Heritage Park and Railway Museum in Lithgow, where Bill Humphreys, Marilu Melo Zurita, and Linda Connor offered further experiential insights into thinking with underground ecologies, as did conversations with all of the diverse participants. The work of the Multispecies Justice (MSJ) Collective, initiated and led by Danielle (Dany) Celermajer at the University of Sydney as part of the Sydney Environment Institute, supported this thinking through its next iteration. Particular thanks to Dany and Sophie Chao, and other participants at various MSJ fora, for their feedback and suggestions in the finalization of this article.

Notes

1. The work of Deborah Wardle introduced me to Darcy's Law: water flow (Q) equals a calculation of permeability of the sand or rock (K), times gravitational head (I), times the area through which the water moves (A). Q = KIA.
2. One of the section headings in Haraway's essay notably underscores "persistence of vision"—yet (as I note at the end of this article) her work here is also to redeem vision as not always necessarily only in service of this master knowledge. However, this otherwise pathbreaking contribution to feminist epistemology and vision does not address the ableism of its pivotal figuration.
3. See the contribution by Neimanis and Chatterjee in Celermajer et al. 2020.
4. In Jamie Linton's analysis, "modern water" emerges from the late eighteenth century onward, with the introduction of the hydrological sciences.
5. Again, for a parallel argument, see Linton 2012.
6. On the nonperformativity of speech acts, see Sara Ahmed 2006. My analysis here is inspired by Ahmed's critique of diversity and inclusion work, whereby the promise to do something can come to stand in for and in fact prohibit taking meaningful further action.
7. See Haraway's "Situated Knowledges" (1988: 595), where she claims that "'objects' do not pre-exist as such." Our knowledge practices shape them in noninnocent ways. When I teach this text I bring a box to the classroom and conduct a number of experiments to see how and by what means we need to shift our perspectives to get to the know the various facets of the box; in the end, as the students discover, the only way to be really sure about what's in the box is to destroy it. Box no more.
8. Celermajer (2006) complicates this reading of Levinas, though, insisting his view of ethical knowledge is more synesthetic, as it is the call of the other that we must also hear in order to act ethically.
9. In addition to their nonsighted bodies, we might also appreciate the temporality of stygofauna as a form of crip time. As described by Mandy Thoo (2012: 35), stygofauna "have no circadian rhythms to clock the time of day as they never see the sun. As a result, compared with their surface equivalents, they grow slowly, have few young, live long lives and stay very close to home. Their slow lifestyle is regarded as an adaptation to the low-energy environment in which they live." In this sense, they might also invite us to desist from "chrononormativity" (Freeman 2010).
10. This resonates with Celermajer's (2006) call for knowledges that are "synesthetic," where knowing via one sense is complemented by others.

References

ABC News. 2017. "WA Government Approves Uranium Project despite Environmental Concerns." January 16. https://www.abc.net.au/news/2017-01-16/wa-government-approves-cameco-uranium-mine/8186514.

Ahmed, Sara. 2006. "The Non-performativity of Anti-racism." *Meridians* 7, no. 1: 104–26.

Ballastero, Andrea. 2019. "Aquifers (or, Hydrolithic Elemental Choreographies)." *Fieldsights*, June 27. https://culanth.org/fieldsights/aquifers-or-hydrolithic-elemental-choreographies.

Bebbington, Andrew. 2012. "Underground Political Ecologies." *Geoforum* 43, no. 6: 1152–62.

Bell, Sarah. 2019. "Experiencing Nature with Sight Impairment: Seeking Freedom from Ableism." *Environment and Planning E: Nature and Space* 2, no. 2: 304–22.

Boulton, Andrew J., et al. 2003. "Imperilled Subsurface Waters in Australia: Biodiversity, Threatening Processes, and Conservation." *Aquatic Ecosystem Health and Management* 6, no. 1: 41–54.

Boulton, Andrew J., and William Humphreys. 2005. "Aquifers and Hyporheic Zones: Towards an Ecological Understanding of Groundwater." *Hydrogeolgoy Journal* 12, no. 1: 98–111. https://doi.org/10.1007/s10040-004-0421-6.

Braidotti, Rosi. 2019. *Posthuman Knowledge*. London: Polity.

Celermajer, Danielle. 2006. "Seeing the Light and Hearing the Call: The Aesthetics of Knowledge and Thought." *Literature and Aesthetics* 16, no. 2: 120–44.

Celermajer, Danielle, et al. 2020. "Justice through a Multispecies Lens." *Contemporary Political Theory* 19, no. 3: 475–512.

Clare, Eli. 2014. "Mediations on Natural Worlds, Disabled Bodies, and a Politics of Cure." In *Material Ecocriticism*, edited by Serenella Iovino and Serpil Opperman, 204–18. Bloomington: Indiana University Press.

Currell, Matthew, et al. 2017. "Problems with the Application of Hydrogeological Science to Regulation of Australian Mining Projects: Carmichael Mine and Doongmabulla Springs." *Journal of Hydrology*, no. 548: 674–82.

de la Cadena, Marisol. 2015. "Uncommoning Nature." *e-flux*, no. 65. https://www.e-flux.com/journal/65/336365/uncommoning-nature/.

Dumit, Joseph. 2014. "Writing the Implosion: Teaching the World One Thing at a Time." *Cultural Anthropology* 29, no. 2: 344–62. https://doi.org/10.14506/ca29.2.09.

Freeman, Elizabeth. 2010. *Time Binds: Queer Temporalities, Queer Histories*. Durham, NC: Duke University Press.

Giraud, Eva. 2021. *Veganism: Politics, Practice, and Theory*. London: Bloomsbury.

Gomez-Barris, Macarena. 2015. "Inverted Visuality: Against the Flow of Extractivism." *Journal of Visual Culture* 15, no. 1: 29–31.

Guzik, Michelle T., et al. 2010. "Is the Australian Subterranean Fauna Uniquely Diverse?" *Invertebrate Systematics* 24, no. 5: 407–18.

Haraway, Donna. 1988. "Situated Knowledges: The Science Question in Feminism and the Privilege of Partial Perspective." *Feminist Studies* 14, no. 3: 575–99.

Hose, Grant C., J. Sreekanth, Olga Barron, and Carmel Pollino. 2015. *Stygofauna in Australian Groundwater Systems: Extent of Knowledge*. Sydney: CSIRO Land and Water/Macquarie University. https://publications.csiro.au/rpr/download?pid=csiro:EP158350&dsid=DS1.

Humphreys, William. 2006. "Aquifers: The Ultimate Groundwater-Dependent Ecosystems." *Australian Journal of Botany* 54, no. 2: 115–32.

Jay, Martin. 1988. "The Rise of Hermeneutics and the Crisis of Ocularcentrism." *Poetics Today* 9, no. 2: 307–26.

Karanovic, Tomislav, Stefan M. Eberhard, Giulia Perina, and Shae Callan. 2012. "Two New Subterranean Ameirids (Crustacea: Copepoda: Harpacticoida) Expose Weaknesses in the Conservation of Short-Range Endemics Threatened by Mining Developments in Western Australia." *Invertebrate Systematics* 27, no. 5: 540–56.

Korbel, Kathryn, et al. 2017. "Wells Provide a Distorted View of Life in the Aquifer: Implications for Sampling, Monitoring, and Assessment of Groundwater Ecosystems." *Scientific Reports*, no. 7: 40702.

Liboiron, Max. 2021. *Pollution Is Colonialism*. Durham, NC: Duke University Press.

Linton, Jamie. 2012. *What Is Water?* Vancouver: University of British Columbia Press.

McLean, Jess, et al. 2018. "Shadow Waters: Making Australian Water Cultures Visible." *Transactions of the Institute of British Geographers* 43, no. 4: 615–29.

Melo Zurita, Marilu. 2020. "Challenging Sub Terra Nullius: A Critical Underground Urbanism Project." *Australian Geographer* 51, no. 3: 269–82.

Melo Zurita, Marilu, Paul Munro, and Donna Houston. 2017. "Un-earthing the Subterranean Anthropocene." *Area* 50, no. 3: 1–8.

Mokany, Karol, Thomas D. Harwood, Stuart A. Halse, and Simon Ferrier. 2018. "Riddles in the Dark: Assessing Diversity Patterns for Cryptic Subterranean Fauna of the Pilbara." *Diversity and Distribution* 25, no. 2: 240–54.

Mudd, Gavin. 2019. "It's Not Worth Wiping Out a Species for the Yeelirrie Uranium Mine." *Conversation*, April 26. https://theconversation.com/its-not-worth-wiping-out-a-species-for-the-yeelirrie-uranium-mine-116059.

Myers, Natasha. 2017. "Ungrid-able Ecologies: Decolonizing the Ecological Sensorium in a Ten-Thousand-Year-Old NaturalCultural Happening." *Catalyst: Feminism, Theory, Technoscience* 3, no. 2: 1–24.

Neimanis, Astrida. 2017. *Bodies of Water: Feminist Posthuman Phenomenology*. London: Bloomsbury.

Poelina, Anna, Kathrine S. Taylor, and Ian Pedrisat. 2019. "Martuwarra Fitzroy River Council: An Indigenous Cultural Approach to Collaborative Water Governance." *Australian Journal of Environmental Management* 26, no. 3: 236–54.

Plumwood, Val. 2008. "Shadow Places and the Politics of Dwelling." *Australian Humanities Review*, no. 44. http://australianhumanitiesreview.org/2008/03/01/shadow-places-and-the-politics-of-dwelling/.

Rasmussen, Claire. 2015. Pleasure, Pain, and Place." In *Critical Animal Geographies*, edited by Kathryn Gillespie and Rosemary-Clare Collard, 54–70. London: Routledge.

Scott, Heidi V. 2008. "Colonialism, Landscape, and the Subterranean." *Geography Compass* 2, no. 6: 1853–69.

Shotwell, Alexis. 2016. *Against Purity: Living Ethically in Compromised Times*. Durham, NC: Duke University Press.

Singh, Julietta. 2018. *Unthinking Mastery: Dehumanism and Decolonial Entanglements*. Durham, NC: Duke University Press.

Thoo, Mandy. 2012. "The Living World below Us." *Australasian Science*, November, 34–36.

Tomlinson, Moya, Andrew J. Boulton, Peter J. Hancock, and Peter G. Cook. 2007. "Deliberate Omission or Unfortunate Oversight: Should Stygofaunal Surveys Be Included in Routine Groundwater Monitoring Programs?" *Hydrogeology Journal* 15, no. 7: 1317–20.

Tschakert, Petra. 2020. "More-than-Human Solidarity and Multispecies Justice in the Climate Crisis." *Environmental Politics* 31, no. 2: 277–96. https://doi.org/10.1080/09644016.2020.1853448.

Wardle, Deborah. 2020. "Storying with Groundwater." *Swamphen*, no. 7: 1–15.

Wesner, Ashley. 2018. "Contested Sonic Space: Settler Territoriality and Sonographic Visualization at Celilo Falls." *Catalyst: Feminism, Theory, Technoscience* 4, no. 2: 1–34.

White, Emma. 2018. "The Limits of Modelling: Knowing What We Don't Know." *Open Forum*, April 22. https://www.openforum.com.au/the-limits-of-modelling-knowing-what-we-dont-know/.

Wright, Alexis. 2006. *Carpentaria*. Artarmon, NSW: Giramondo.

Yusoff, Kathryn. 2018. *A Billon Black Anthropocenes or None*. Minneapolis: University of Minnesota Press.

Zakharia, Nickolas. 2020. "Mining Industry Holds Largest Slice of Australian Economy." *Australian Mining*, November 2. https://www.australianmining.com.au/news/mining-industry-holds-largest-slice-of-australian-economy/.

Astrida Neimanis writes about water, bodies, and weather from intersectional feminist perspectives. They are currently associate professor and Canada Research Chair in Feminist Environmental Humanities at the University of British Columbia Okanagan, where they are also director of the FEELed Lab. Astrida is a white settler from the Baltic Sea region who grew up mostly on Anishinaabe and Haudenosaunee lands in Southern Ontario.

UNEARTHING the TIME/SPACE/MATTER of MULTISPECIES JUSTICE

Christine J. Winter

Abstract Multispecies justice is a developing field—or perhaps more accurately, a set of fields. It draws together a range of academic disciplines to examine human and nonhuman relationships. These include relationships of respect, responsibility, and, to some, reciprocity. The extent of those relationships and the range of species, forms, and being to be included, however, remains indistinct and variable. Whereas within traditional theories of justice concern for other beings remains tied to the desire to enhance human experience, life opportunities, goods, and virtues, the call to multispecies justice is motivated by the recognition that the nonhuman realm has intrinsic value and values. This article's argument is that given the relative infancy of multispecies justice as a field of study in the Western academy, there is an opportunity to ensure that it examines not only how to avoid damaging domination of the nonhuman realm but also the ongoing colonial domination of Indigenous epistemologies and ontologies. The article does not suggest an appropriation of Indigenous knowledge but rather an exploration of ways in which the field may remain sufficiently nuanced and open to accommodate multiple epistemological and ontological framings of theory. Drawing from Mātauranga Māori the article discusses an aspect of that decolonial project—why the scope of multispecies justice needs to be open to all planetary being and all time.

Keywords multispecies justice, decolonial, Mātauranga Māori, space-time, matter

The summer of 2019–20 was one of anxiety in Australia. For months across the southeastern seaboard the sky was an ocher canvas painted with the ashes of incinerated

Cultural Politics, Volume 19, Issue 1, © 2023 Duke University Press
DOI: 10.1215/17432197-10232459

flora and fauna. That ocher ash didn't stay in the sky; it insinuated itself onto our decks, over our windowsills, and into our lungs. We, the humans and nonhumans of Australia, have imbibed that dead-life. We are one with it. It is one with us. This was, as my colleague Danielle Celermajer (2020) describes it, the season of omnicide—the killing of all things. A time when the slow violence (Nixon 2011) of capitalism and the Anthropocene burst across our screens, into our lives, at furious pace. Graphic images of conflagration caught the public eye. How could they not? Was something new shimmering in the public consciousness? While the firefighters and government focused on property and human life, as they are quite rightly charged with doing,[1] public attention seemed to shimmy between human and nonhuman suffering and loss.[2] The sight of Australia's charismatic kangaroo and koala burned and distressed was too much to bear. Chris Dickman of the University of Sydney calculated that over three billion vertebrates had lost their lives in the inferno (Readfern and Morton 2020). Over one-third of Australia's native forests were burned to cinders (Werner and Lyons 2020). Many more dead critters and organisms have gone uncounted: fungi, worms, algae, mycorrhizae, unseen contributors to the planetary system. As I read the stories, watched video footage, and listened to conversations, I asked myself if I was witnessing something new, the birthing of an embryonic sense of multispecies justice within the public imagination. Perhaps this was simply a projection of my own sympathies and inclination, but media reports and social media were flooded with pictures of fleeing or flayed animals with backdrops of raw roaring flames or silent scenes of blackened earth and skeleton stakes of once-towering trees.[3]

Whether or not I am right to think the media and community are thinking beyond accustomed ways of motivating care for the environment and the nonhuman realm, there *is* a growing recognition that neither climate justice nor environmental justice theories gives sufficient voice to the nonhuman realm as subjects of justice in their own right (see esp. Celermajer, Chatterjee, et al. 2020; Celermajer, Schlosberg, et al. 2020; Tschakert et al. 2021). While concerned for the environment, including plant and animal life, most variants of climate and environmental justice remain resolutely anthropocentric. Multispecies justice (MSJ), on the other hand, centers no one species, human or otherwise. MSJ recognizes that the nonhuman death and damage wrought by events like those I began with are more than simply regrettable, tragic, or lamentable. They are matters of justice. And that claim cracks open the scope of who is the subject of justice. This leads me to present here a self-consciously ontological politics of MSJ. A politics that respects difference. One that is sufficiently capacious, fluid, and nuanced to accommodate a pluriverse of ontologies (Escobar 2015, 2020).[4] One that can enfold the human, nonhuman, and spiritual realms.

However, in thinking about the subject of MSJ, the Western academy confronts some strongly held notions about the significance of species and human exceptionalism, and therefore who counts ethically and politically before the law. That is, political theorists, environmental philosophers, and others encounter various culturally bound ontological certainties: ontological certainties that have their roots deeply embedded in Western philosophy; ontological certainties, derived from the Anglo-European philosophical tradition, that make it particularly hard to theorize justice for the nonhuman and to move beyond compassion or sympathy for the multiple forms of being and beings that create the earth system.

In the Western academy MSJ is a developing field—or perhaps more accurately, fields. It draws together a range of academic disciplines to examine human and nonhuman relationships. These include relationships of respect, responsibility, and, to some, reciprocity. The extent of those relationships and the range of species, forms, and being remains indistinct and variable. Hence this article's intervention and suggestion that this fluidity and variability are essential if MSJ is to avoid the universalizing instincts of Western philosophy and politics that Indigenous peoples, for instance, find exclusionary. If the modes of including multiple species into the spheres of justice operate only within narrow epistemological and ontological frames, MSJ could exclude and dominate, and direct violence toward, Indigenous and other peoples whose lives are bound within different philosophical frames. That would perpetrate and perpetuate injustice.

In the context of contemporary debates about the moral status of the nonhuman, some identify the relationships and bounds of responsibility as purely anthropocentric—motivated by the desire to enhance human experience, life opportunities, goods, and virtues.[5] For others the nonhuman realm has intrinsic value and values, and this intrinsic worth motivates the call for MSJ.[6] My argument here is that given the relative infancy of MSJ as a field of study *in the Western academy*, there is an opportunity to ensure it examines not only how to avoid damaging domination of the nonhuman realm but also the ongoing (neo)colonial domination of Indigenous epistemologies and ontologies. I am not suggesting an appropriation of Indigenous knowledges but rather an exploration of ways in which the field may remain sufficiently fluid and open to accommodate multiple epistemological and ontological framings of theory. I am suggesting we are present to an imminent moment—a moment in which there is potential for what emerges from the encounter between West and Indigenous to be transformative, a time/space "in which different ways of being and doing find interesting things to do together" (Rose 2017: G51). In this article I will draw from Mātauranga Māori to discuss one aspect of such a project—why the scope of MSJ needs to accommodate all planetary being within an expansive, nonmechanical, nonlinear conception of time/space/matter. I am suggesting MSJ can and should walk in a companionable time/space/matter of multiple knowledges and multiple ontologies (human and nonhuman), ensuring it is, to return to Arturo Escobar, pluriversal at its very heart.

I write this as a citizen of two countries where Indigenous peoples' sovereign claims remain insufficiently recognized by the contemporary settler-colonial state. I am a descendant of Anglo-Celtic settlers and Ngati Kahungunu of Aotearoa New Zealand, where I was born and raised. And I am a settler in Australia, where I live, work, and play on unceded Gadigal lands. My motivations to explore the importance of decolonizing theory in the settler states[7] is based in part in this personal positionality. They also emerge from the politics and ethics that arise from ongoing issues of sovereignty and the denigration and domination of Indigenous peoples' being, knowledge, philosophies, cultures, and laws. More generally, I am challenging any claims to universality, which, if not explicitly made, are at least assumed within academic theories of justice. I am not positing another "universal." Rather, I aim to highlight an opportunity in which MSJ opens theory, in a nondogmatic way, to inclusion. I suggest all time and all matter can matter in the sense that the boundaries of MSJ's description of "species" needs to be nonexclusive. Porous boundaries might then accommodate multiple ontologies without hierarchical ordering or domination.

My starting point is general. In section 1 is a claim that decolonization of theory is a matter of justice and theories that foreclose Indigenous epistemologies and ontologies are unjust. This is followed by two sections in which I focus firstly on what matter matters—or to whom justice is applicable in a decolonized and capacious theory of MSJ. That discussion then draws our attention to *wātea* (time/space) and the ways in which both theory and politics/policy imagine the time/space for which they are accountable. I conclude with a provocation to those working in the field of MSJ to examine the foundations of their thinking and the breadth of its application by proposing that a conception of a subject of MSJ must be sufficiently open to accommodate all time/space/matter. That is, I am not saying that all time/space/matter must always be included but that the potential for their inclusion is not foreclosed by the theory. My claim is that it is timely for the MSJ community to step into a space of scholarship that neither transcends nor dominates the Indigenous "other" but walks alongside the long intellectual traditions of First Nations peoples. To rise together in the shimmer, the shimmer of the biosphere and ancestral life that Deborah Bird Rose (2017: G53) describes coming from an aesthetic response, an "appeal to the senses"—that shimmer I felt (or imagined) emanating from the wider community through the long horror of the black summer of 2019–20. A horror made more urgent by the knowledge that it was a harbinger of more to come; made still more poignant by the knowledge that the First Peoples of the continent have MSJ practices, protocols, and procedures that the overlay of Western politics and law forbids them enacting. If, in imagining MSJ, theorists embrace a process of encounter and transformation, not absorption, the humanities might enter the zone of the shimmer. They may "produc[e] . . . new, immanent modes of existence" (Stengers 2010: 35) in which their partners are Western, Indigenous, multispecies, and more. The theory may then be thought of as pluriversal, open, and receptive to multiple ontologies meshing in multiple (and maybe messy) ways across the globe while operating at local levels. A resultant state of ontological fluidity and capaciousness may unfold. I offer here two inflection points to achieve such an outcome—the matter of matter and the matter of wātea—time/space/matter.[8]

1. Why Does Decoloniality Matter?

Eve Tuck and Wayne Yang (2012) argue that there will be no "decolonization" until the land is returned to the Indigenous peoples of the settler states. For settlers, people encoded with an understanding that land is property/wealth, such a notion may send chills through their very being. They will lose wealth and the Indigenous peoples gain it. This appears unfair. Of course, it is a simple reversal of the processes of colonialism, processes that have never ceased, as Patrick Wolfe (2006) so clearly outlined. Citizens of the settler states may fear the return of property from the majority to a minority would tip the scales in the direction of new and unwarranted inequality—the many would become the poorer and the (undeserving) Native immensely wealthy.

Ironically, perhaps, it is not the desire for monetary wealth, nor punishment, nor retribution that drives many Indigenous people to call for the return of their lands. The driving motivation is multifold: to reestablish relationships of care for and reciprocity with the land, waters, seas, critters, and plants of their world; to reclaim language and culture; to revive cosmologies; and thus, to reclaim identity. It is, too, to seek recognition of the ongoing significance and efficacy of Indigenous knowledge, science, philosophy, and culture in caring for multispecies relationships and welfare—in caring for land, waters, creatures, and plants. Indigenous cultural ambivalence to property as wealth signals a complex set of relationships in the twenty-first-century settler states. Indigenous people do "own" property, and some are wealth builders in the capitalist sense of the word. But it is not this alone that drives the call for restitution of lands. A different sort of wealth is sought. A wealth that comes from caring for and carefully tending to the health and well-being of human, nonhuman, and spiritual realms and an understanding that each realm is dependent on the well being of the others for its well-being. The desire is to be whole again. Put simply, decolonization will only be complete when the lands are returned to Indigenous care. Decolonization is not about including some Indigenous history in the syllabus, or reciting a statement of recognition, or even fully including Indigenous people in the creation of research projects. The settler states will not be decolonized until Indigenous peoples are once more the custodians of the land and fully reconnected with it. Settler states will not be decolonized until, as in the other global colonies, sovereignty is returned to the original peoples of the lands.[9] However, until then academics can acknowledge Indigenous knowledge and philosophy fully rather than (albeit perhaps unconsciously) perpetuating colonial knowledge practices. MSJ must become self-consciously critical of the potential violence of the epistemological and ontological foundations of theory and practice.

Before I proceed to consider how these principles speak to the development of MSJ as an academic field, I want to stake a critically important claim. There is nothing "new" about the field of MSJ. It is important for political theorists, people interested in justice and particularly those on the cusp of "creating" a subfield called MSJ to remember this is not a new field of theory. It is a field of philosophy that has engaged the minds of Indigenous peoples (and possible all peoples) for millennia. It is a field of philosophy that is well theorized and operationalized. Indigenous peoples have thought and continue to think deeply about these matters; have developed complex understandings of the interweave of human and nonhuman; have protocols, procedures, and practices to ensure the resulting ideas of justice are enacted; and provide ongoing evidence of their efficacy. It is important, therefore, that academic theorists register three things as they "develop" this subbranch of theory called MSJ. First, and to reiterate, it must never be prefaced by the word "new"—it is old, theorized and practiced by cultures around the world, just not by the Anglo-European sphere. Second, it is important to credit Indigenous thinkers, to acknowledge the prior work of generations of Indigenous knowledge makers. It is not okay to do as so many of the "progressive" thinkers of the "new" posthumanist school have done and simply ignore that extant knowledge and theory, to pretend they have been untouched by its subtle influence on their own thinking (Todd 2016). Decolonization also means citing Indigenous scholars, crediting their knowledge, and acknowledging their influence. It means accepting Indigenous knowledge holders as intellectual peers and producers of knowledge, and not (just) as research subjects. And finally—and it is to this point this article speaks—it is imperative that the theory that is "created" is ontologically fluid and open, that it is capacious enough to hold multiple philosophical heritages and influences, that it can rest comfortably within and nurture a pluriverse. What is required of the ontological foundations of MSJ to ensure it evades the trap of sustaining the colonial project? How can a theory of MSJ ensure it does not continue to eliminate the Native (Wolfe 2006)?

Many have noted the problems of Enlightenment/post-Enlightenment dualities. The problems are highlighted in environmental justice (EJ) (Schlosberg 2007), intergenerational justice (De-Shalit 1995; Hiskes 2005, 2009), climate change justice (Gardiner 2011), deep green ecology (Leopold 1968), and earth justice law (Bosslemann 2011), and by environmentalists. The idea that human and nonhuman are independent of each other, that the nonhuman realm is granted to humans to dominate, abuse, and debase at will and without repercussion grows from this dualistic thinking and is seen as the root of many environmental woes. It is understood, too, by critical EJ scholars to be at the root of the racism that underscores many environmental injustices (Bullard 1994: Pellow 2016), and ecofeminists highlight not only its destructive but also its patriarchal impulses (Braidotti 2016; Macgregor 2017; Neimanis 2017; Plumwood 2000, 2002). Despite this recognition, environmental, intergenerational, and climate justice and early attempts to bring nonhuman into the spheres of justice (Nussbaum 2007) remain anthropocentric. This is problematic not only for the other beings with whom humans share the planet but also for Indigenous theorists. It is problematic because such a distinction is anathema in Indigenous philosophy (Alfred 1999, 2005; Coulthard 2014; Kimmerer 2013; Watene 2016; Winter 2019; Yunkaporta 2019). It is problematic, too, for finding a path through the problems of justice these theories are attempting to resolve. This is the first of the ontological cleavages that MSJ must address. Arguably it does. The point of MSJ is to bring the nonhuman realm into the spheres of justice and, in so doing, make it a subject of justice. This, however, leads to the questions the next section will address: What matter matters? Who/what is accepted as subjects of justice? Are there limits to being justice-worthy? What is the unit of consideration? How to ensure that who/what is defined as a subject of justice does not foreclose Indigenous peoples' understanding of being and standing?

Within the question of what matter matters are matters of scale. Spatial scale is implied within the question itself. What is the smallest unit to which justice applies? What is the largest? Where are the boundaries for the theory placed? Indeed, are they necessary? How are boundaries justified? To ensure the theory is decolonial, the scale of who/what matters must not automatically exclude some forms of being; animal, vegetable, mineral, and spiritual, as I will explain shortly.[10] Beyond the anthropocentric, and beyond demanding that only those others that have capacities that resemble human capacities fall within the circle of justice, questions of spatial scale emerge. Second, what is the temporal scale? The question here is, Is a mechanical, forward-looking, progressive understanding of time sufficient to encompass Indigenous ontologies? If not, can MSJ acknowledge multiple temporalities?

2. *What Matter Matters?*

Considering what matter matters starts by engaging liberal philosophy's attachment to individualism. I will not examine individualism in detail here—other MSJ scholars do that satisfactorily (see Celermajer, Schlosberg, et al. 2020; Schlosberg 2012), and critiques abound in multiple Indigenous scholarships (Coulthard 2014; Durie 2010; Graham 2008, 2013). What is important here is that the scale of justice cannot simply be the individual if MSJ is to respond to and, importantly, not dominate Indigenous ontologies. While the individual continues to matter and continues to be a subject of justice, it is not all that matters, and it is not the sole unit of consideration (Durie 2010; Marsden 2003; Mead 2003). For instance, within Mātauranga Māori—Māori knowledge traditions—the focus is not on the individual being or species but on the relationships that generate well-being (Marsden 2003; Roberts 2010). One has responsibilities to others as a function of one's responsibility to one's self. As we support others, human and nonhuman, and foster their well-being and growth, so we develop our own standing and worth. These practices of relating and relationality are contained within *tikanga* (the protocols and practices of Māori ethics and philosophies), and more specifically in *manaatikanga* (the practices of providing support, care, hospitality, and generosity and showing respect for others), and *whanaungatanga* (connectedness to self; family and others; places; and the multispecies realm secured through relationships) (Mead 2003; Williams 2001). Subjectivity is, then, relational, and relationality generates responsibility not in the sense of liberal rationality but rather in *whanaungatanga* and *manaatikanga*. It is our relationship to all things that is the precursor of MSJ. The subject of justice, then, in Rose's terms, emerges from the (preconscious) shimmer that can be felt in/from others, that exists first, and in which human's responsibilities are prefigured.

Māori philosophy and knowledge tell us the individual matters as a component within a nested web of relationships between human and human, human and nonhuman, nonhuman and nonhuman, nonhuman and human. The framework of Māori epistemology—*whakapapa*—itself means to place in layers, or to lay one on top of another.[11] "For something to exist and be known, it must have a whakapapa; put another way, in order to 'know' about a thing (or a person) one must know its whakapapa" (Roberts 2012: 743). We must know its relationships. All matter has whakapapa. It is an epistemology that layers through commonalities, connections, and interconnections. Importantly, "Whakapapa act as cognitive maps, delineating spatial and temporal discontinuities within a particular environment" (743–44). This relational ontology links human, nonhuman, plant and animal, and form and spirit in the present from the past and into the future. Relationships and interconnectivity can "contain a variety of what modern science would classify as living and non-living entities—a distinction which does not apply to the Māori worldview" (745). Coming from an oral tradition, whakapapa might be seen as spoken maps or genealogical charts that relate (or layer) species with places and seasons, ancestors and gods, and responsibilities and moral codes. The upshot of this account of the world and knowledge structure is that nothing (and no time) is considered in isolation. In forming our knowledge of a particular plant species, for instance, that knowledge is "laid down with" "spatial, temporal and functional interactions between [that plant] and its surroundings" (745). In this way it is possible to understand relationships between living and nonliving, such that that no unit can be understood without its contextual relationships. Nothing is left "alone" and individualized.

What then becomes apparent is that in a multispecies community each component has a valuable contribution to make to the whole, and it is the protection and nurturing of *relationships* and *relationality* that is the primary concern. It means, too, that the power humans have within these relationships fashions a set of *specific* responsibilities to the multispecies community. A set of responsibilities to support and enhance the flourishing of the community for all time. It is when those responsibilities are abrogated or denied as a function of dualism and political and/or religious authorities, which claim the nonhuman realm is the domain of human domination, that balance within the natural realm is undone. Others have shown that thinking through a dualist lens legitimates the racialized distribution of environmental harms (Bullard 1994; Pellow 2016). A decolonized ontology recognizes the individual can only be strong, healthy, and fully functioning where the whole is thriving. This is not to dismiss justice to the individual as an existing unit; it is to say the individual only exists and may only experience justice within the larger network of relational responsibilities and functions and so the subject of justice is *more than* the individual, more than human. For decolonial MSJ to do justice to a set of relationships, it might conceive that justice resides in the relationship, not the individual or species. Furthermore, the "species" of multi*species* justice cannot be circumscribed. The sets of relationships might extend between animal, vegetable, and mineral, living and nonliving, sensate and insensate, from individual being to the multitudes held within the unity of a landscape.

One of the strengths of using *multi* and *species* in MSJ is that each term holds open a space for an individual, a unit composed of many, and the planetary system, simultaneously—if MSJ theory allows it to. The multi(plicity) of MSJ opens the potential of justice theory beyond environmental justice's parameters, which so often focus on the distribution of harms to one unit (be that an individual person or community of people) within an oftentimes proscribed domain. It is receptive to pluriversal world views in a way distributional theories of justice are less adept at accommodating (and might therefore be open to accusations of injustice inherent in the theories parameters). A receptivity to including all matter in the realms of justice draws us to a theory of "encounter and transform[ation] not absorption" (Rose 2017: G51). And that seems compatible with the purpose of MSJ; if the theory's purpose is to expand the spheres of justice beyond the human, it should, I suggest, simultaneously be ontologically accessible to all human knowledge practices. The question then is one of defining the boundaries of a theory of local webs of relationships within a web of planetary relationality.

One recurrent focus of Indigenous philosophies of MSJ is reciprocity. Reciprocity exists at numerous scales within the natural world and is also a matter of human moral and ethical responsibility. For instance, in forests,

> the mycorrhizae may form fungal bridges between individual trees, so that all the trees in a forest are connected. These fungal networks

> appear to redistribute the wealth of carbohydrates from tree to tree. A kind of Robin Hood, they take from the rich and give to the poor so that all the trees arrive at the same carbon surplus at the same time. They weave a web of reciprocity, of giving and taking. In this way, the trees all act as one because the fungi have connected them. Through unity, survival. All flourishing is mutual. Soil, fungus, tree, squirrel, boy—all are the beneficiaries of reciprocity. (Kimmerer 2013: 20)

The emerging "scientific" understanding that Robin Wall Kimmerer (an academic biologist and member of the Citizen Potawatomi Nation) explains here is situated at a meeting point with long held Indigenous knowledge, and MSJ is, I suggest, on the cusp of such a meeting point. Importantly, recognizing human responsibilities of reciprocity to the nonhuman realm can be an element of MSJ. One of the sets of responsibilities one takes on as an Indigenous person is an obligation to reciprocate for the gifts given by the natural world to human flourishing. That is done through respect for and nurturing of that realm.[12]

Practically, this means MSJ expands well beyond, say, the reduction of the suffering of sentient animals, as Martha Nussbaum and Rachel Wichert have argued is a matter of justice (Nussbaum 2007; Wichert and Nussbaum 2017). It is this and it is more. If MSJ is framed in a way that allows for altering landscape through, say, the redirection of some water from a watercourse—something that may be deemed a minimally invasive intervention by government and policy makers—or more controversially, perhaps, does not include the water and watercourse as subjects of justice, then it excludes Indigenous understandings of what it is to be in the world, and to whom/what one owes responsibilities of reciprocity.[13] Responsibility for the flourishing of the habitat from which the watercourse was redirected is part of it and not all. There is a responsibility to the integrity of the whole; to the water and the watercourse and the habitat and the critters and plants, fungi and mycorrhizae, and to the spirit of the place—the wonder that sits somewhere at the limits of our ken, shimmering at the edges of our understanding—and the spirit of creation. MSJ must be *sufficiently capacious* to account for justice to the living, nonliving, and spiritual—to support pluriverses—if it is to avoid the violence of dominating Indigenous ontologies.

When thinking about what matter matters, what are scholars working at the cusp of a westernized MSJ choosing to exclude? What are the criteria? How will decisions about the validity of exclusion (Barad 2007), if any, be made? As noted earlier, the very term *species* in MSJ implies only living things; however, Mātauranga Māori—Māori knowledge—tells us *all* matter matters. It tells us too that size is not important. It tells us the individual exists only within a web of supportive relationships, and a component of those relationships may be as small as a microbe harnessing an even more minute speck of mineral that will in time be passed up the trunk of a mighty kauri tree to support its structure, growth, and thriving. All matter matters because all matter interlinks, locally and on a planetary scale. How can justice be framed to acknowledge nested being and nested responsibilities to reciprocal relations between living and nonliving, sentient and nonsentient, human and nonhuman, human, nonhuman and spiritual? How are the "waves of ancestral power that shimmer and grab [that] are also exactly the relationships that bring us forth and sustain us" (Rose 2017:

G61) captured? It is a matter of decolonial justice that MSJ eschews the universalist impulse and embraces the powerful potential, the liberation, of openings for multiple philosophical traditions.

3. The Matter of Time and Space

If all matter matters, does all space matter? Once all matter matters, perhaps all time must matter also. Where the subject of justice is an individual human person, the spatial and temporal scales are limited by the physical boundaries and life span of that identified individual. This is part of Derek Parfit's (1984) argument in his examination of intergenerational justice. Beyond the lifespan of an identifiable human, the subject to whom justice applies cannot be identified, and so those who fall outside these time space boundaries are left outside the boundaries of theory and obligation. This is, says Parfit, a "repugnant conclusion," as it limits intergenerational justice to overlapping generations, while matters of climate and environmental justice extend well beyond those concurrent intergenerational boundaries. The scalar problem MSJ now confronts is that the subject of justice is something more than the individual; perhaps it is a community of like beings, or perhaps communities of multiplicitous relationships. Maybe it is the configuration of "spacetimematter"[14] that Karen Barad (2007: 244) encourages in (re)evaluations of "questions of boundary, connectivity, interiority, and exteriority (topographical concerns)." There are sub-subjects within that community that exist for mere seconds (such as a raindrop) through to those that date from seconds after the big bang—the elements that make matter, including hydrogen, which makes up 10 percent of the human body. The construction of *wātea*—time/space—the way MSJ imagines it might expand to something more than a continuum that relegates the past to some immaterial realm, prioritizes the present, and discounts the future while simultaneously valorizing "progress."

Within a Māori understanding, space and time are one. Referred to as *wātea*, time/space are "conjoined together and relative to each other" (Marsden 2003: 61). This means that in any place, time past, present, and future are caught in a web of relationships with that space, the people of the place, and the nonhuman and spiritual relationships embedded in and associated with it. Merata Kawharu (2010: 222) explains it this way:

> Māori have a particular way of dealing with the present. There is a well known aphorism that says Māori walk backwards into the future, that is, they take the past with them in advancing into the unknown. Present and future circumstances are made sense of by referencing the past and therefore all contained within it—ancestors, gods and spiritual powers. Past, present and future are collapsed into one. Interpreting "environment" then, is not simply about considering a place or landscape in the present, but also about taking into account times past or history, and all that it contains.

Within the environment are markers, signs, and aide-mémoire that hold the history of *iwi* or *hapu* (tribal and subtribal groups), that bring the ancestors and the spirits to life in the present. The whakapapa of the iwi or hapu holds these layers—ancestors, gods, and the present—together within the landscape or environment. What this means, when all matter matters, is that responsibilities to the nonhuman realm also encompass responsibilities to ancestors, future generations, and the spiritual realm. So MSJ does not focus on any one species or time period or space but rather cocoons them within an ongoing one—time/space/matter. Each iwi and hapu's identity and each Māori person's identity are referenced through key features of the landscape in which they are embedded, and to which they hold responsibilities. Ancestors and time are blended within the "marae locale"—the territory of the iwi/hapu. "'Environment' has meaning not only in economic sustainability terms, where resources provide, for instance, the basic food and clothing necessities, but in political and cultural/spiritual terms as well. All these elements are strongly intertwined" (Kawhara 2010: 222). Kawharu writes that the Marae (the formal gathering place and meeting house of hapu and iwi) represents "environment" for Māori. The building itself is a representation of time/space/matter of earth, sky, ancestors—human and spiritual, and the world between. The environment of concern to Māori is that of their *rohe*, or territory—their "locale." Thus, when implementing justice within the environment, the protocols and practices are locally based. However, because the underlying philosophies governing the practices are common while the focus may be local, the outcome is a coherent environmental and species protection regime across the (is)lands and seas. The scale of environmental responsibility may be local but the fundamental foundation of philosophy, whakapapa—the entangled human/nonhuman/ spiritual—links across iwi and the islands of Aotearoa. The local is connected to all time/space/matter through whakapapa, and hence so are one's responsibilities to the environment—to MSJ, as I am suggesting here. How might a theory of MSJ respond to a local/present/all space/all time/all matter interweaving spiral? A spiral that moves both inward and outward. In attempting to respond to that question, I revert to the practical/physical.

The living species of MSJ can be very long lived. Think of a one-thousand-year-old kauri tree (*Agathis australis*) or three-thousand-year-old olive (*Oleo europaea*) which may have a further three thousand years of life ahead. Or of the eighty-thousand-year-old quaking aspen (*Populous tremuloides*) clonal colony in south-central Utah. How can MSJ "do" justice to those pasts and to their future? Let's think of the less spectacular forest tree. One with a less dramatic life span. What temporal scale is required for them? Superficially it might equate to a human span, and that is not all. For the tree is fed not only by the ancient minerals of the soils beneath but also by its decomposing self—the aged branches and leaves it itself drops to decompose at its base. And simultaneously at its base are the seeds it shed last season, cosseted under the leaf mold, tucked away, awaiting the rains or maybe more light to themselves put down roots and thrust forth leaves and continue the cycle of life. The temporality here reaches from the ancient and decomposing to the living and thence to a potentiality of new life.

A political theory of MSJ should not foreclose thinking beyond spans that reflect human life spans. A flexible foundation will accommodate conceptions of continuous spirals of life cycle accommodating the very long lived and ever regenerating. How is that achieved while also accommodating the very short-lived, the insect, for instance, whose life may span a mere day? It's difficult within an imaginary that conceives of time as a forward moving projection. Forward-moving point-to-point projections provide a useful framing of time for some purposes. However, it is just one imaginary, within which this moment exists and then is past. Yet the moment I inhabit now was a moment ago the future—that is, the moment "now" is an infinitesimal instant held momentarily in existence, wedged between the past it is soon to be and the future it no longer is; time is always twisting within a past present future spiral, this instant referenced within the past and future. I am suggesting MSJ can think of time in much thicker terms than as a simple forward projection, and any conceptualization of MSJ needs to be open to that possibility if we are to accommodate the multitemporal being of the nonhuman realm. It is also required of a decolonized MSJ.

I've argued elsewhere that time colonizes intergenerational justice theory (Winter 2020), and I argue here it is important to avoid the same misstep when conceiving MSJ. The strictly linear projection that grounds the dominant Western imaginary of time is not universal.[15] For instance, for Indigenous peoples in Australia, time is understood as circular—circular in the way I just described for the tree living from its ancestor, with its seed nurtured at its own base by the leaves that fall from its branches before it too collapses, allowing the seedling to feed from it as it reaches into the newly light-filled void (Lee 2006; Povinelli 2016; Watson 2015). Underscored by this temporal circularity, the obligations of justice are responsible simultaneously to the past, present, and far future. This does two things relevant for my argument. It expands the subject beyond the currently living individual (human, animal, plant, rock formation, etc.), first backward to include responsibilities to ancestors/antecedents, and second, forward to include future generations of its and other kinds of being, while also accommodating justice in the present. Thus the set of responsibilities is complex and comprehensive—and much more demanding than those conceived within linear temporal bounds. A simple contractual arrangement between living sentient beings cannot deal with this. There is something much subtler at play here.

Māori understand time within a spiral construct. At any point within the temporal spiral one references backward and forward. So while the moment exists, while a lifetime exists, the moment and the life are referenced back to ancestors—to all that is laid down within the layers of whakapapa—and that which will in future be incorporated within the same construct. "Mountains, ranges, lakes and waterways, and all living matter, human included, have a repetitive pattern, a pattern of birth, change and rebirth of continuity, circularity and synchrony" (Winter 2020: 289), and this needs to be accommodated within conceptualizations of MSJ.

That is all rather abstract. How might this idea of time, of a spiraling swirl in which the present has before it the past and future, apply to MSJ? Let's this time consider a river. A river, such as the Whanganui in Aotearoa, which is recognized by the contemporary state as a living identity, one that has interests, rights, powers, and duties (New Zealand Government 2017). This status is a legal identity status.[16] That the Whanganui is the holder of interests and rights suggests that it is a subject of justice. How can MSJ accommodate their temporalities?

In Mātauranga Māori a river is not only water tumbling, sliding, drifting between banks of ferns and mosses. A river is more than this. A river is the banks, ferns, and mosses, a river is its bed of eroded rubble, of bedrock and silt. A river is its relationships—with the mountain snows, the glacial melt, the freshwater spring, the drizzling mists and pounding rains that feed it. A river is its tributary rivulets, the surrounding forests, the fish within, the tuna (eel) and inanaga (*Galaxias maculatus*), the dragonfly and midge that dance above. And the river is its relationship with its human custodians—the river is the people and they the river—as this Whanganui iwi statement of belonging and identity reflects:

E rere kau mai te Awanui
Mai i te Kahui Maunga ki Tangaroa
Ko au te Awa, ko te Awa ko au

The Great River flows
From the Mountains to the Sea
I am the River, and the River is me (New Zealand Government 2017)

This cascading multiplicity of relations exists within a multiplicity of temporalities. The timeless bedrock, which converts to rubble and silt to be swept to sea, deposited to be compressed and consolidated back into rock forms, must be accommodated along with the droplets of drizzle born of the morning mists that kiss the stream, becoming one with the larger body of water. It must account for the ancient trees of the forest and the budding fern at its edges. This array of relations exists within multiple temporalities that MSJ needs to account for. The time/space/matter of the leaf wafting to join the mat of leaf mold that generates the quintessential fragrance of petrichor. A decolonial, pluriversal MSJ must capaciously understand that "within Māori ontological and cosmological paradigms, it is impossible to conceive of the present and future as separate and distinct from the past, for the past is constitutive of the present and, as such, is inherently reconstituted within the future" (Stewart-Harawira 2005: 42).

4. Unearthing the Time/Space/Matter of MSJ

Relationships between people and the natural environment, between tangible and intangible dimensions, between organic and inorganic material, and between past and future constitute the foundations upon which indigenous populations understand the world.
—Mason Durie, "Outstanding Universal Value: How Relevant Is Indigeneity?"

In shaping this article, I stated that others had dealt with the issue of individualism and I was not going to. However, in the course of this examination of scale—the scale of the subject of justice and the temporality of MSJ—I have demonstrated (obliquely perhaps) that to linger within an imaginary of the individual is impossible in a comprehensive and decolonized theory. To do so limits imagination to a fascination with the power of what it is to be human, and human in a very particular and limiting way. That in turn constrains the boundaries of MSJ and its potential, limiting the good that MSJ can achieve. Moreover, it excludes the intellectual and philosophic traditions of Māori—once more casting "the Indigenous" to the margins of what is understood as reason. Justice then becomes unjust.

That the subject of MSJ is something more than an individual permeated the discussion of what matter matters—which I argue is all matter, as none may exist without the other. A tree can grow to its fullness only within a set of entangled relationships with fungi and mycorrhizae, its own leaf litter, and possibly in association with its own rotting mother-tree at its base and with seeds awaiting its own death to spring forth. Each of these relationships also relies on the extraction by the fungi of minerals from rock and soil, sustaining the interplay of nutrients and sugars, photosynthesis and respiration. Justice for the tree or the bird plucking fruit from its branches and scattering its seed depends on relationships with the entire complex. If MSJ is described such as to exclude any element of the complex, creating an artifice of separation, it rends the whole. To dismiss or discount a component is eventually to do an injustice to the whole and all elements within that whole.

To step away from speciesism is necessary if MSJ is to eschew the (unjustly dominating) colonial intellectual tradition. To achieve that, it must create space for all time/space/matter. That is, it must adopt openness as an ontological starting point. In the settler states particularly, this is important as they weave toward an inclusive politics, one that addresses the constant presence of oppressive and unjust processes of colonialism. Framing MSJ more radically and capaciously is one way in which theory can promote and support the decolonizing ambitions of the First Nations peoples of the settler states. I suggested as a first prerequisite that a theory of MSJ must recognize preexisting Indigenous philosophies in which there are well-developed theories of MSJ, and which furthermore have effective protocols, practices, and procedures for implementing the theory. That is, MSJ may never be prefaced with "new," for it is not, and to do so is to dismiss millennia of intellectual and philosophical enterprise, enterprises that continue into the present.

More precisely here, the argument has been that all matter can matter within decolonized imaginaries of MSJ, because all matter matters within (at least some) Indigenous philosophies. In Mātauranga Māori, for instance, the intellectual construct, the epistemological foundations of Māori philosophy and science, depend on whakapapa—the setting down in layers of the living, nonliving, ancestral, and spiritual in dependent relationships. These interdependencies, the search for commonality and relationships, become the basis for MSJ. A recognition, if you like, that abuse of an element is an abuse of the whole. This philosophical foundation is common to each Māori hapu and iwi (the primary political units of Māori society); the locale of implementation is local.[17]

And that is not all; while physical scale may be local, wātea/timespace is expansive. All time matters. A decolonial MSJ will be able to respect and do justice to the ancestors, the living and future generations—human and nonhuman, living and elemental, material and spiritual. MSJ in a decolonialized register will recognize and respect the gifts of the past—be they human ancestors' care of the environment or fertile and productive soils formed by forests and deep-rooting grasses, rotting critters and dung, the insects, microbes, fungi, and algae that generate that fertility. MSJ will not negate the possibility that past lives within the present as much as the present inhabits the future and accounts for relationships of justice to that tangled spiral of time and being(s). MSJ might shimmer in the "brilliance of motion and encounter" (Rose 2017: G53), if all matter can matter through all wātea/timespace.

Notes

1. Thousands of homes were lost, 33 lives lost, 11.3 million people affected by smoke, and 10.6 million concerned about their own and others' safety (Werner and Lyons 2020); an estimated 445 others died later from effects of the fires (Wahlquist 2020).
2. This observation is derived from the prominence of photographs on animal suffering in news reports and social media feeds.
3. See, for example, Australian Academy of Science 2020; *Climate Signals* 2021; Slezak, 2020.

4. The idea of the pluriverse is one that Arturo Escobar champions as a means of escaping the domination of neoliberal politics and capitalism that erase alternative ontologies in the bid to privatize and capitalize the nonhuman realm. The pluriverse questions the "belief in a single reality" (Escobar 2020: 3) and opens the way to a world of "possibilities" that may reorient people and governments toward a less self-destructive mode of being.
5. This is the focus of much environmental justice, for instance.
6. For a comprehensive review of the state of MSJ and approaches to it, see Celermajer, Schlosberg, et al. 2020.
7. The settler states are Aotearoa New Zealand, Australia, Canada, and the United States. Previous colonies of the British, they are politically differentiated from the colonies of Africa and Asia for instance for three reasons: the British (and other Europeans) removed the original inhabitants from the land—by force and "legal" fiat—and distributed it to themselves. That is, they came to stay; they imposed their own system of law and governance over the whole population, initially denying Indigenous enfranchisement; and the settlers and settler descendants significantly outnumber the Indigenous population, that is the Indigenous populations are minorities (Aotearoa Māori 16.5 percent, Australia 3.3 percent, USA 1.7 percent, Canada 4.9 percent of their countries' total populations).
8. There are other points of inflection we might seek, but I restrict the discussion to just two in this article. On reciprocity, see, for example, Whyte and Cuomo 2017; Kimmerer 2013. On dignity, see Watene 2016; Winter 2019.
9. Another distinguishing feature of the settler states is that sovereignty was not returned to the original Peoples as occurred elsewhere in the great decolonial push of the 1950s and 1960s. Initiated by the United Nations, colonizing nations were required to relinquish rule in "their" colonies, which were then supported toward self-determination and self-government. The settler states were distinguished from these other colonies because they were no longer directly governed by the colonial power—that is, they were self-governing, with their own governments, constitutions and legal systems—and therefore they were not required to return sovereignty to the First Peoples/First Nations. Return of sovereignty, then, is, as Eve Tuck and Wayne Yang (2012) identify, the overriding ambition of decolonization.
10. Here we run into a question about the word *species* in MSJ. From a Māori perspective this may not be the most suitable word, because there is no distinction made between animate and inanimate within the epistemological and ontological frameworks of Māori (see Roberts 2010, 2013; and Roberts et al. 2004). However, *multispecies justice* is the coinage, and to quibble seems both churlish and to undercut the good work MSJ seeks to achieve in the world. Rather, it seems to me better to take a capacious view of the meaning of species—to understand it as a placeholder for the multiple types of being/matter, animate and inanimate, that make up the planetary system.
11. *Whakapapa* is frequently translated as "genealogy," a recitation of relationships reaching back to creation and framing the possibilities of future.
12. Robin Wall Kimmerer (2013) writes particularly well on this theme. For an Australian Indigenous view, see also Graham 1999, 2008, 2013. For a comprehensive examination of Aotearoa New Zealand Māori philosophy, see Roberts 1998; Roberts et al. 2012.
13. There is a long history of Māori protests and delegations resisting the diversion of rivers for hydroelectric power generation in Aotearoa. These include the redirection of waters from the headwaters of the Whanganui River—a matter than has not been fully resolved by the legislation granting the Whanganui legal identity status—and multiple dams across the mighty Waikato resisted by Tainui as *kaitiaki* of the river.
14. From here on I use *time/space/matter*—not to snub Barad's formation, nor to differentiate for the sake of it, but rather because the Māori term *wātea* is made from combining *time* (*wā*) and *space* (*atea*).
15. It has, of course, been universalized by globalization in many contexts.
16. This status has been negotiated between iwi—whose traditional rohe (territory) lies within

the area and whose claims to territoriality include responsibilities to care for the multispecies-natural and spiritual realms—and the government. It is a blending of the iwi's understanding of the ontological status of the Whanganui River and Anglo/New Zealand legal structures. It need not be conceived as an anthropocentric formulation but rather as a blend of the "one" Whanganui—that is, a multiplicity of elements human, nonhuman, spiritual, ancestor, living, and future. This is not to say that in other jurisdictions where this ontology does not apply, "personhood" status might continue the anthropocentric domination of nonhuman realms.

17. Thanks to Chris Crews and Lisa Ellis for drawing my attention to the importance of scale of implementation.

References

ABC730 (@abc730). 2019. "Lewis, the koala taken to the Port Macquarie Koala Hospital for treatment of injuries suffered in the NSW bushfires, has not survived." Twitter, November 27. https://twitter.com/abc730/status/1199630410465959936.

Alfred, Taiaiaki. 1999. *Peace, Power, Righteousness*. Don Mills, ON: Oxford University Press.

Alfred, Taiaiaki. 2005. *Wasáse: Indigenous Pathways of Action and Freedom*. Toronto: University of Toronto Press.

Australian Academy of Science. 2020. "At Least a Billion Animals Killed in Bushfires." YouTube video, 3:33. January 9. https://www.youtube.com/watch?v=dx1WnWjo-yQ.

Barad, Karen. 2007. *Meeting the Universe Halfway: Quantum Physics and the Entanglement of Matter and Meaning*. Durham, NC: Duke University Press.

Best, Daniel (@20thCenturyDan). 2021. "Never, ever forget the silent victims of #Bushfires." Twitter, January 25. https://twitter.com/20thCenturyDan/status/1353868027847004161.

Bosselmann, Karl. 2011. "From Reductionist Environmental Law to Sustainable Law." In *Exploring Wild Law*, edited by Peter Burdon, 204–13. Kent Town, SA: Wakefield.

Braidotti, Rosie. 2016. "The Critical Posthumanities; or, Is Medianatures to Naturecultures as *Zoe* Is to *Bios*?" *Cultural Politics* 12, no. 3: 380–90. https://doi.org/10.1215/17432197-3648930.

Bullard, Robert D. 1994. "Environmental Justice for All." In *Unequal Protection: Environmental Justice and Communities of Color*, edited by Robert D. Bullard, 3–22. San Francisco: Sierra Club Books.

Celermajer, Danielle. 2020. "Omnicide: Who Is Responsible for the Gravest of All Crimes?" Religion and Ethics. *ABC Religion and Ethics*, January 2. https://www.abc.net.au/religion/danielle-celermajer-omnicide-gravest-of-all-crimes/11838534.

Celermajer, Danielle, Sria Chatterjee, Alasdair Cochrane, Stefanie Fishel, Astrida Neimanis, Anne O'Brien, Susan Reid, K rithika Srinivasan, David Schlosberg, and Anik Waldow. 2020. "Justice through a Multispecies Lens." *Contemporary Political Theory* 19, no. 3: 475–512. https://doi.org/10.1057/s41296-020-00386-5.

Celermajer, Danielle, David Schlosberg, Lauren Rickards, Makere Stewart-Harawira, Mathias Thaler, Petra Tschakert, Blanche Verlie, and Christine Winter. 2020. "Multispecies Justice: Theories, Challenges, and a Research Agenda for Environmental Politics." *Environmental Politics* 30, nos. 1–2: 119–40. https://doi.org/10.1080/09644016.2020.1827608.

Climate Signals. 2021. "Australia Bushfire Season 2019–2020." Last updated October 15. https://www.climatesignals.org/events/australia-bushfire-season-2019-2020.

Coulthard, Glenn. 2014. *Red Skin, White Masks*. Minneapolis: University of Minnesota Press.

De-Shalit, Avner. 1995. *Why Posterity Matters*. New York: Routledge.

Durie, Mason. 2010. "Outstanding Universal Value: How Relevant Is Indigeneity?" In *Māori and the Environment*, edited by Rachel Selby, Pātaka Moore, and Malcolm Mulholland, 239–49. Wellington: Huia.

Escobar, Arturo. 2015. "Commons in the Pluriverse." In *Patterns of Commoning*, edited by David Bollier and Silke Helfrich. Common Strategies Group.

Escobar, Arturo. 2020. *Pluriversal Politics: The Real and the Possible*. Durham, NC: Duke University Press. https://doi.org/10.1215/9781478012108.

Gardiner, Stephen M. 2011. *A Perfect Moral Storm: The Ethical Tragedy of Climate Change*. Oxford: Oxford University Press. http://doi.org/10.1093/acprof:oso/9780195379440.001.0001.

Graham, Mary. 1999. "Some Thoughts about the Philosophical Underpinnings of Aboriginal Worldviews." *Worldviews* 3, no. 2: 105–18.

Graham, Mary. 2008. "Some Thoughts about the Philosophical Underpinnings of Aboriginal Worldviews." *Australian Humanities Review*, no. 45: 181–94.

Graham, Mary. 2013. "Custodial Navigator." *Colourise*. http://colourise.com.au/wp-content/uploads/2015/08/Custodial_Navigator1.pdf (accessed November 19, 2022).

Hiskes, Richard P. 2005. The Right to a Green Future: Human Rights, Environmentalism, and Intergenerational Justice. *Human Rights Quarterly* 27, no. 4: 1346–64. http://doi.org/10.1353/hrq.2005.0049.

Hiskes, Richard P. 2009. *The Human Right to a Green Future: Environmental Rights and Intergenerational Justice*. Cambridge: Cambridge University Press.

Kawharu, Mereta. 2010. "Environment as a Marae Locale." In *Māori and the Environment*, edited by Rachel Selby, Pātaka Moore, and Malcolm Mulholland, 221–37. Wellington: Huia.

Kimmerer, Robin. W. 2013. *Braiding Sweetgrass: Indigenous Wisdom, Scientific Knowledge, and the Teachings of Plants*. Minneapolis: Milkweed Editions.

Lee, Martin, and Bob Randall. 2006. *Kanyini*. Directed by Melanie Hogan. https://www.vacca.org/page/get-involved/cultural-hub/video/kanyini-documentary-pg.

Leopold, Aldo. 1968. *A Sand Country Almanac*. Oxford: Oxford University Press.

Macgregor, Sherilyn, ed. 2017. *Routledge Handbook of Gender and Environment*. London: Routledge.

Marsden, Māori. 2003. *The Woven Universe: Selected Writings of Rev. Māori Marsden*. Edited by Te Ahukaramū Charles Royal. Otaki: Estate of Rev. Māori Marsden.

Mead, Hirini. M. 2003. *Tikanga Māori*. Wellington: Huia.

Neimanis, Astrida. 2017. *Bodies of Water: Posthuman Feminist Phenomenology*. London: Bloomsbury Academic. http://doi.org/10.5040/9781474275415.ch-001.

New Zealand Government. 2017. *Te Awa Tupua (Whanganui River Claims Settlement) Act 2017*. http://www.legislation.govt.nz/act/public/2017/0007/latest/whole.html.

Nixon, Rob. 2011. *Slow Violence and the Environmentalism of the Poor*. Cambridge, MA: Harvard University Press.

Nussbaum, Martha. C. 2007. *Frontiers of Justice*. Cambridge, MA: Belknap Press of Harvard University Press.

Parfit, Derek. 1984. *Reasons and Persons*. Oxford: Clarendon.

Pellow, David. N. 2016. "Toward a Critical Environmental Justice Studies: Black Lives Matter as an Environmental Justice Challenge." *Du Bois Review* 13, no. 2: 221–36. http://doi.org/10.1017/S1742058X1600014X.

Plumwood, Val. 2000. "Integrating Ethical Frameworks for Animals, Humans, and Nature: A Critical Feminist Eco-socialist Analysis." *Ethics and the Environment* 5, no. 2: 285–322.

Plumwood, Val. 2002. "Decolonising Relationships with Nature." *PAN*, no. 2: 7–20.

Povinelli, Elizabeth. A. 2016. *Geontologies: A Requiem to Late Liberalism*. Durham, NC: Duke University Press.

Readfern, Graham, and Adam Morton. 2020. "Almost Three Billion Animals Affected by Australian Bushfires, Report Shows." *Guardian*, July 28. https://www.theguardian.com/environment/2020/jul/28/almost-3-billion-animals-affected-by-australian-megafires-report-shows-aoe.

Roberts, Mere. 2012. "Mind Maps of the Māori." *GeoJournal* 77, no. 6: 741–51. http://doi.org/10.1007/s10708-010-9383-5.

Roberts, Mere, Bradford Haami, Richard Anthony Benton, Terre Satterfield, Melissa L. Finucane, Mark Henare, and Manuka Henare. 2004. "Whakapapa as a Māori Mental Construct: Some Implications for the Debate over Genetic Modification of Organisms." *Contemporary Pacific* 16, no. 1: 1–28. http://doi.org/10.1353/cp.2004.0026.

Rose, Deborah Bird. 2017. "Shimmer: When All You Love Is Being Trashed." In *Arts of Living on a Damaged Planet: Ghosts and Monsters of the Anthropocene*, edited by Anna Tsing, Nils Bubandt, Elaine Gan, and Heather Swanson, 52–63. Minneapolis, University of Minnesota Press.

Schlosberg, David. 2007. "Defining Environmental Justice: Theories, Movements, and Nature." Oxford: Oxford University Press.

Schlosberg, David. 2012. "Justice, Ecological Integrity, and Climate Change." In *Ethical Adaptation to Climate Change: Human Virtues of the Future*, edited by Allen Thompson and Jeremy Bendik-Keymer, 165–83. Cambridge, MA: MIT Press.

Slezak, Michael. 2020. "Three Billion Animals Killed or Displaced in Black Summer Bushfires, Study Estimates." ABC News, July 27. https://www.abc.net.au/news/2020-07-28/3-billion-animals-killed-displaced-in-fires-wwf-study/12497976.

Stengers, Isabelle. 2010. *Cosmopolitics*. Vol. 1. Translated by Robert Bononno. Minneapolis: University of Minnesota Press.

Stewart-Harawira, Makere. 2005. *The New Imperial Order.* Wellington: Huia.

Todd, Zoe. 2016. "An Indigenous Feminist's Take on the Ontological Turn: 'Ontology' Is Just Another Word For Colonialism." *Journal of Historical Sociology* 29, no. 1: 4–22. https://doi.org/10.1111/johs.12124.

Tschakert, Petra, David Schlosberg, Danniel Celermajer, Lauren Rickards, Christine Winter, Mathias Thaler, Makere Stewart-Harawira, and Blanche Verlie. 2021. "Multispecies Justice: Climate-Just Futures with, for, and beyond Humans." *WIREs Climate Change* 12, no. 2: e699.

Tuck, Eve, and K. Wayne Yang. 2012. "Decolonization Is Not a Metaphor." *Decolonization* 1, no. 1: 1–40.

Wahlquist, Calla. 2020. "Australia's Summer Bushfire Smoke Killed 445 and Put Thousands in Hospital, Inquiry Hears." *Guardian*, May 26. https://www.theguardian.com/australia-news/2020/may/26/australias-summer-bushfire-smoke-killed-445-and-put-thousands-in-hospital-inquiry-hears.

Watene, Krushil. 2016. "Valuing Nature: Māori Philosophy and the Capability Approach." *Oxford Development Studies* 44, no. 3: 287–96. http://doi.org/10.1080/13600818.2015.1124077.

Watson, Irene. 2015. *Aboriginal Peoples, Colonialism, and International Law.* New York: Routledge.

Werner, Joel, and Suzannah. Lyons. 2020. "The Size of Australia's Bushfire Crisis Captured in Five Big Numbers." *ABC Science*, March 4. https://www.abc.net.au/news/science/2020-03-05/bushfire-crisis-five-big-numbers/12007716.

Whyte, Kyle P., and Chris J. Cuomo. 2017. "Ethics of Caring in Environmental Ethics: Indigenous and Feminist Philosophies." In *The Oxford Handbook of Environmental Ethics*, edited by Stephen M. Gardiner and Allen Thompson, 234–47. Oxford: Oxford University Press.

Wichert, Rachel N., and Martha C. Nussbaum. 2017. "Scientific Whaling? The Scientific Research Exception and the Future of the International Whaling Commission." *Journal of Human Development and Capabilities* 18, no. 3: 356–59. https://www.tandfonline.com/doi/full/10.1080/19452829.2017.1342386.

Winter, Christine J. 2019. "Decolonising Dignity for Inclusive Democracy." *Environmental Values* 28, no. 1: 9–30. https://doi.org/10.3197/096327119X15445433913550.

Winter, Christine J. 2020. "Does Time Colonise Intergenerational Environmental Justice Theory?" *Environmental Politics* 29, no. 2: 278–96.

Williams, David. 2001. *Matauranga Māori and Taonga.* Wellington: Waitangi Tribunal.

Wolfe, Patrick. 2006. "Settler Colonialism and the Elimination of the Native." *Journal of Genocide Research* 8, no. 4: 387–409. https://doi.org/10.1080/14623520601056240.

Yong, Ed. 2016. *I Contain Multitudes*. London: Harper Collins.

Yunkaporta, Tyson. 2019. *Sand Talk: How Indigenous Thinking Can Save the World.* Melbourne: Text.

Christine J. Winter (Ngati Kahungunu ki Wairoa, Ngati Pakeha) is a senior lecturer at the politics program at the University of Otago/Te Whare Wānanga o te Ōtākou in Aotearoa New Zealand and a research associate at the Sydney Environment Institute. Her research focuses on the ways in which academic political theory, and particularly theories of justice, continue to perpetuate injustice for some people (and more specifically for Māori) and the environment. Her most recent research centers on ensuring the emerging field of a political theory of multispecies justice should have decolonial (and anticolonial) foundations.

TRANSITIONAL JUSTICE beyond the HUMAN

Indigenous Cosmopolitics and the Realm of Law in Colombia

Daniel Ruiz-Serna

Abstract **Indigenous and Afro-Colombian peoples often describe the harm caused by armed conflict in terms of damage inflicted on their traditional territories. To these peoples, the concept of territory makes reference not only to their lands but to a set of emplaced practices and relationships through which they share life with wider assemblages of human and other-than-human beings. It is the threat faced by these large communities of life that was invoked by Indigenous organizations when they succeeded in including the territory as a victim in the transitional justice framework recently implemented by the Colombian state. This article argues that the consideration of the territory as a victim means more than the full enjoyment of the land ownership rights Indigenous and Afro-Colombian peoples are entitled to. Instead, said consideration challenges some received notions regarding justice and reparation, particularly because war becomes an experience that extends beyond human losses and environmental degradation. The terms and practices mobilized by Indigenous and Afro-Colombian peoples compel us to examine the limits that concepts such as human rights, reparation, or even damage have in the understanding of war and its aftermath.**

Keywords **territory, warfare, Indigenous and Afro-Colombian peoples, transitional justice**

War is a disastrous scourge that has hit Indigenous peoples and Afro-Colombian rural communities the hardest (United Nations 2004, 2010; Forst 2018). Besides the violations of their fundamental human rights and the damage

Cultural Politics, Volume 19, Issue 1, © 2023 Duke University Press
DOI: 10.1215/17432197-10232473

inflicted on their unique cultures, some of the most flagrant harms perpetrated against these peoples and their traditional territories include land conversion, natural resource depletion, ecosystem pollution, and biodiversity loss (Centro de Memoria Histórica 2009; Franco and Restrepo 2011; J. Vargas 2016). In order to tackle some of these issues and as part of a suite of policies that paved the way for the recognition of rights for all victims of armed conflict, the Colombian president signed the Victims and Land Restitution Law (Law 1448) in 2011. It aimed to provide reparations to victims of human rights violations and to restore the stolen lands of the millions of peasant families that had been driven away from their homes during the last decades of armed conflict. Although the law was a major step toward addressing the legacy of violence, it was strongly criticized by Indigenous and Afro-Colombian social movements because the government did not seek their prior and informed consent during its formulation.[1] Additionally, these grassroots organizations argued that, given the fact that the armed conflict has affected Indigenous and Afro-Colombian peoples disproportionately, the damage to their unique ways of being required different measures of rectification than those the government provided to other victims.

It is in this context of Indigenous organizations pressing for recognition of their experience that the state and Afro-Colombian and Indigenous social movements forged legal measures specifically intended to redress the damage inflicted on Indigenous territories. Concretely, the Decree-Law 4633, also known as the victims' law for Indigenous peoples, incorporates the idea that, alongside human beings and ways of life, traditional territories should also be considered victims of war: "The territory, understood as a living entity and foundation of identity and harmony, in accordance with the very cosmovision of the indigenous peoples and by virtue of the special and collective link that they hold with it, suffers damage when it is violated or desecrated by the internal armed conflict and its underlying and related factors."[2] Recognizing the importance of traditional lands as the material condition for the lives these peoples lead, and considering "the special spiritual relationship . . . that Indigenous peoples have with their territory" (Art. 8), the Decree-Law confers legal personhood to Indigenous territories and provides an important starting point in the consideration of the long-lasting effects of war on the lands that Indigenous communities inhabit. In this article, I argue that the inclusion of territory as a victim opens a plethora of possibilities for mending wrongs of war that go well beyond considerations of human or environmental damage. This is particularly important given that Indigenous peoples consider that their territories encompass different ontological domains, peopled by different kinds of other-than-human beings who are endowed with agency and personhood. Using a political ontology approach that draws on the work of anthropologist Mario Blaser (2009a, 2009b, 2013, 2016), I show that the recognition of Indigenous territories as victims should mean more than simply the full enjoyment of the land ownership rights that these peoples are entitled to. Rather, what is at stake here are the very relations through which human and other-than-human beings sustain their mutually constituted lives and render the territory "a living entity," in the parlance of the law. In order to stress my point, I bring to the fore a distinction between *territorial damage*—all those actions that, like land dispossession, depletion of natural resources, or forced displacement, limit the effective enjoyment of ownership rights—and *damage to territory*—those actions that jeopardize the relationships that communities cultivate with the myriad other-than-human beings that constitute their territories. This distinction, I argue, allows us to see the victims' law for Indigenous peoples as a unique opportunity to decolonize justice and decenter the human in our understanding of war and its aftermath.

Based on ethnographic work conducted over a period of sixteen years in Bajo Atrato, northwest Colombia, and relying on ongoing collaboration with Indigenous and Afro-Colombian social organizations in this region, I discuss how the recognition of a form of damage that extends to a variety of other-than-human beings poses a series of challenges to the cultural policies promoted under the banner of multiculturalism as well as to the implementation of appropriate and effective policies of truth, justice, and reparation for these peoples. In the first part, I offer a brief overview of how Indigenous and Afro-Colombian peoples conceptualize territory and show how their ideas of justice are always embedded within their territories. I then describe the world-making significance of war and the importance of Indigenous experiences in assessing damage and envisaging reparations. In the third part, I demonstrate how the victims' law might propel a cosmopolitical understanding of more-than-human relationships. I close by identifying some of the challenges that the recognition of Indigenous territories as victims poses to a multicultural state engaged in redressing the harm provoked by armed conflict.

But first, let me unpack two of the main concepts in the analysis that follows. Initially coined by Isabelle Stengers (1996) and then reinterpreted by Bruno Latour (2004b), the notion of cosmopolitics refers to the ways in which sociohistorical arrangements open up space for other-than-human entities within the cosmos to compose a common, political world (Stengers 2005a). Whereas Stengers characterizes this common world as "always aspired to and never achieved" (Watson 2011: 73) and the political task of building it as a matter of indetermination rather than conciliation, Latour sees said common world as the outcome of a progressive composition in which things and other-than-human beings gather together alongside their human representatives. In both cases, the core idea is that, through a constant arrangement of contingent associations, humans and other-than-humans construct reality together and forge a collective society or *polis*. Here I engage with Latour's understanding and see cosmopolitics as a form of politics that constantly makes and unmakes the fissure between subjects and objects, nature and society, or facts and values, giving meaning to and shaping the set of practices that sustain a given world or ontology (Blaser 2009b). In this usage, cosmopolitics is connected to a second notion: political ontology. This is a conceptual framework developed by Mario Blaser (2009a, 2009b) that accounts for the conflicts and negotiations that take place when different worlds or ontologies meet within a given power-laden structure and strive to maintain their own existence and framework of intelligibility amid inter-worldly encounters and frictions. A brief example drawing from discussions about the effects of war and reparation policies in Bajo Atrato helps illustrate how these concepts are intertwined.

In Colombian legal framings, discussions of reparation focus mainly on economic and material means to ensure the return of Afro-Colombian communities to the lands they were expelled from by guerrilla and paramilitary armies. In contrast, some communities in Bajo Atrato express other kinds of concerns as they manifest their intention *not* to return to these lands because they are now haunted by the spirits of those who suffered a violent death and whose bodies were never buried. Discussions around the effects of war and reparation policies thus assume an ontological form: two different ways of experiencing territory are enacted, informing the kinds of actions people think ought to be undertaken and the beings that should be involved to guarantee the return of people to their lands. An inquiry based on the concept of political ontology thus provides the conditions under which the actions and beings mobilized by different actors, including the state, achieve a degree of intelligibility within a given political domain, including the terms for the design of adequate reparation policies.

Warp and Woof of a Life Lived Otherwise

Colombian jurisprudence has made Indigenous territories the cornerstone of the protection of these peoples, since territory is considered "the material possibility for ethnic groups to exercise their rights to cultural identity and autonomy, insofar as this is the physical space in which their culture can survive."[3] Legally speaking, territory might be considered a fundamental right: Indigenous and Afro-Colombian communities enjoy imprescriptible and inalienable collective land tenure. When understood as "material possibility" or "physical space," territory evokes the terms of a substantialist ontology in that it comprises the assortment of places a given community inhabits as well as the set of resources that community uses for its livelihood and to which, only a posteriori, communities come to attach social and cultural meaning. However, it is worth mentioning that this use always takes place in particular historical and cultural contexts embedded within the wider array of acts, institutions, and relations that make up a given society. This means that even in its substantial or material dimension (i.e., the physical enclosure and the diversity of natural landscapes, resources, and species that constitute a place), territory does not precede Indigenous peoples. By this I mean that territory is not something that is complete in and of itself, nor is it separate from or prior to the values, practices, and relations that render places and lives meaningful.

Territory is at the heart of the political struggles Indigenous and Afro-Colombian social movements have undertaken around the defense of their ethnic identities and cultural rights. In the 1970s, for example, when developmentalist policies in Colombia aimed to increase rural productivity by transforming the lands the state considered idle (i.e., lands that belonged to Indigenous peoples), Indigenous organizations argued that what they were defending was not so much land (*tierra* in Spanish) as territory (*territorio*), meaning that the defense of their lives involved not just the protection of their property rights but the recognition of the knowledges, values, and collective modes of being cultivated in existential places (Escobar 2008). Unlike some definitions underscoring territory as a bounded spatial entity (Delaney 2005) and territoriality as attempts to assert or enforce control over specific geographic areas (Sack 1983), Indigenous and Afro-Colombian coneptualizations of territory are akin to phenomenological and socio-constructivist approaches to place (i.e., territory is not a neutral ground, it stands at the core of the very experience of human beings as agents, and it is what results from everyday place-bound social practices; see Casey 1996; Ingold 2000; Lefebvre 1991). Also similar to the way Gilles Deleuze and Félix Guattari (2004) conceptualize territory as a livable order produced and sustained by personal or collective arrangements, the notion of territory has come to incorporate, for Indigenous and Afro-Colombian peoples, different but interconnected objectives: the securing of sustainable livelihoods and collective land ownership, conservation of local knowledge and traditional practices of production, defense of traditional cultures, the experience of place and a deep sense of belonging, proper forms of governability, or the autonomy to make decisions about development policies (Escobar 2008). What is at stake, then, when defending traditional territories is a form of being, place, and life that challenges modern assumptions regarding land ownership and property rights (Boyd 2017; Escobar 2015).

During the last three decades, armed conflict and the rapacious economic policies of consecutive Colombian neoliberal governments have seriously hindered Indigenous and Afro-Colombian communities' access to territory. Some of the long-lasting effects of war on ethnic territories include forced displacement, the transformation of land use, the plundering of natural resources, massacres, loss of crops and biodiversity, extinction, poverty, and dispossession, to name but a few. And yet, for people from Bajo Atrato, above and beyond the aforementioned effects, war also causes fundamental damage to the territory when, for example, it provokes the disappearance of the spiritual masters that protect game animals; the anger of sylvan spirits that, in turn, brings about new kinds of diseases; or the madness of snakes and other poisonous animals. In short, as I later explain, the territory is harmed when human and other-than-human beings are prevented from maintaining the practices and relationships that help them sustain their worlds and their emplaced sense of being. This is why territories for Indigenous and Afro-Colombian peoples constitute not only a space of life but also a living entity. Let me contextualize this statement.

When I first arrived in Bajo Atrato in 2003, the region was the scene of a major humanitarian crisis provoked by the war waged by guerrilla, paramilitary, and military armies for the subjugation of its population and the control of the cocaine trafficking corridor linking Colombia to the huge North American market. Thousands of Afro-Colombian families had been driven from their lands, and hundreds of Indigenous communities were confined to their villages. This critical situation led local grassroots organizations to work with national and international NGOs, alongside the most progressive wing of the Catholic church, to find safe strategies for the return of communities to their lands and to take a nonviolent stand against armed actors. At the time, I worked with one of these NGOs, advocating the implementation of legal suits against the Colombian state for its financial support to companies colluding with paramilitary armies that had violently seized collective lands and converted them into vast oil palm plantations. In 2005, on the verge of a crucial legal decision regarding the rights of ownership over these lands, one of the most charismatic local leaders, Father Armando Valencia—an Afro-Colombian priest whose work was inspired by the liberation theology and whose political stance inspired local communities and rookies like myself—delineated some key aspects to take into consideration when defending Indigenous and Afro-Colombian territories:

> Territory is the space appropriated for our physical, social, and cultural production. It is the physical space, the plants and the animals; it is the space we name, use, walk, and travel. It is the form villages and households are emplaced, the economy, the ways of living and working, the days for cultural and religious celebrations, the social relationships, our traditional authorities, and our worldview. All these actions unfold in space and they create territoriality, which in turn, helps build what territory is made of. . . . The territory is a space to produce life and culture, it reflects our worldview. In the fields we work, in the social and family relations we keep, in the symbolic aspects of our thinking, the territory is materialized. . . . Territory is not only land because it extends far beyond the physical space granted by the law. (Valencia 2005: 15–20)

This powerfully evocative definition describes key ontological aspects of more-than-human relationships within traditional territories. First, social practices and relationships (e.g., "ways of working," "cultural and religious celebrations," "traditional authorities," and "social and family relations") do not simply unfold in territory. Rather, they contribute to the very creation of territory. Second, territory and communities are mutually linked and reciprocally constituted: diverse practices express the attributes of particular places and the territory itself reflects the qualities of its inhabitants ("In the fields we work, in the social and family relations we keep, in the symbolic aspects of our thinking, the territory is materialized"). Third, territory cannot be understood in abstraction from the experience of being Afro-Colombian and belonging to an Afro-Colombian rural community ("It is the space we name, use, walk, and travel"). This is a sophisticated conceptualization that underscores the way territory participates essentially and not just contingently in the generation of a collective sense of being, how it provides a particular placement to social experiences, and, most importantly to the point being made, how territory does not necessarily exist prior to the relations and practices that take place on and with it. In other words, Father Valencia's definition is anchored in and expressive of a relational ontological approach: territory does not precede the relations that constitute it (Escobar 2016) but is instead produced performatively (Barad 2003). Rather than serving simply as a setting for the lives, human and other-than-human, unfolding in its forests and rivers, territory is an ensemble of relationships and beings that emerge together in a place experienced by Afro-Colombian and Indigenous peoples as an "existential space of self-reference" (Escobar 2003: 53) but also, I would add, as a space of alter-referential encounters, given that humans, other-than-humans, and territory constitute each other's possibilities of existence.

From this conceptualization, it follows that an understanding of war cannot be attained without taking into consideration the effects that armed violence has had on the wider assemblages of human and other-than-human entities. This demands a kind of ethnographic attunement that neither limits itself to a description of the intimate relations between Indigenous peoples and the natural world nor simply includes animals or the environment in its accounts of armed conflict and its aftermath. What is required instead is a consideration of the violent effects of war in terms that are not restricted to humans or their purported universal rights. This is true for two main reasons. On the one hand, the human rights framework is still too human-centric; that is to say, it presumes a neat cut between humans and the material environments with which they are entangled (Boyd 2017). On the other hand, it draws on atomistic and homogenizing conceptions of the human being (i.e., the only authorized subject of law is the autonomous, free individual) that fail to take into consideration the lively relationalities through which humans, places, and other-than-humans are mutually constituted and emerge as large communities of life (Schlosberg 2019). These lively relationalities, as I explore in the next section, come to the fore when considering the harm experienced by the kind of entities that constitute Indigenous and Afro-Colombian territories.

War and Its Afterlives

According to some elders from Bajo Atrato, the constant presence of soldiers in their region has infuriated *los dueños de los animals* (the spiritual masters that protect and release certain game animals) and consequently compromised their food security. Other elders maintain that their lands have been polluted by the unburied bodies of those who suffered terrible deaths. These haunted spirits are also said to frighten away families that finally managed to return to their former villages after several years of forced displacement. Still others say that evil spirits, known in the Emberá language as *andomía*—a term I roughly translate as "water mother"—have wreaked havoc since being released by powerful shamans in their attempt to protect communities from the raids carried out by guerrilla and paramilitary armies. These examples bring to the fore the way armed violence is an experience shared by the myriad beings that Indigenous and Afro-Colombian territories harbor. The disappearance of spiritual protectors of game, the deterioration of habitable lands, and the difficulties in containing the forest's evil forces correspond then to a series of wounds that affect not only the human rights of these peoples but also, and to an even greater degree, the web of embedded relationships and attendant responsibilities these communities enter into with animals, spirits, plants, ancestors, and places.

The far-reaching effects of armed conflict on assemblages of beings and the world-making relations these beings undertake brings into question whether the framework of human rights and its human-centrism are sufficient in the quest of Indigenous peoples for justice and reparation within their traditional territories. War has without doubt compromised human and environmental rights. But something in the way Indigenous peoples experience the damage of war exceeds these rights-based frameworks because the myriad beings affected stand in a relation to the domains of nature and culture in a way that cannot be qualified in terms of either "and" or "or" (de la Cadena 2015). What is at stake is the entangled existence of humans and other-than-humans whose possibilities of being and becoming are inherently and a priori relational. To redress the harm provoked by war in these territories demands that we resist conceptualizing humans and their environments as two distinct ontological domains. This stance has been pushed for in different ways by Indigenous scholars (see, for example, Todd 2018) as well as in different strands of thought about the more-than-human question (Latour 2004a; Ogden, Hall, and Tanita 2013; Viveiros de Castro 1998). What I show is that these ideals must also be embedded into socio-legal frameworks.

In the context of an Indigenous cosmopolitical endeavor that involves sociality between human and other-than-human actors, the inclusion of territory as a victim paves the way for new contestations about what constitutes damage and what forms of reparation are appropriate. Everything related to the restitution of rights to the enjoyment of property speaks about the rights of these communities over their land and resources—and what undermines these rights is what counts as territorial damage. But from the cosmopolitical perspective that Indigenous and Afro-Colombian communities represent, the inclusion of the territory as a victim speaks rather—or also—to the well-being of the territory itself and the beings that comprise it. The actions that compromise this well-being constitute what locals experience as damage to territory. This nuanced yet important distinction calls into question the terms that a multicultural state uses to interpret harm, either locating it in the "real" (what would be described as damage to nature) or in the "cultural" (what would be described as damage to the representations that certain peoples have of nature). Two short examples might illustrate the importance of the distinction between territorial damage and damage to the territory in the design of reparations.

Snakes and Tramas

In the rainforest of Bajo Atrato, home to about 5 percent of all reptiles worldwide (Proyecto Biopacífico 1998; Rangel 2015), venomous snakes are no trivial concern. They are the most feared creature and are considered to be the principal threat to humans. Some people maintain that while deforestation has forced certain animals to retreat, this has not been the case for snakes. In fact, places along the Curvaradó River have seen an increase in snake populations as a result of forest clearing to make way for oil palm plantations by companies in criminal alliances with paramilitary groups. According to the communities that returned to these lands after several years of forced displacement, snakes now represent a major danger because their bites have become more lethal than ever.

Under certain circumstances, a snakebite might become an even more delicate matter than usual. For instance, for those who had sexual intercourse the night before being bitten, the venom might prove lethal. If one has been bitten and needs to cross a river, the venom will act faster. Or, if you managed to kill the snake that bit you but it dies faceup, your treatment will be more complicated. These types of events are considered a *trama*, something that renders one's recovery more difficult. The word *trama* might simultaneously be translated as both woven thread and as plot, in the sense of a storyline. A *trama* amplifies the effects of the poison, making any healing treatment more difficult by somehow twisting or deviating a snakebite's possible outcomes. One can also *tramar* oneself if one does not follow the restrictions prescribed by traditional healers. Finally, some older snakes may have a venom that proves to be stronger than usual, while others may become more venomous because they have bitten several dogs or people. In those cases, their bites easily become *tramadas*, so that the possibility of *tramar* might also be a feature acquired by particular snakes.

Since 2005, communities inhabiting the Curvaradó River have noticed an increase in the snake population and in the bites being *tramadas*, which means that these snakebites are difficult to heal with the treatments normally used by traditional healers. This increase is associated with the presence of large swaths of oil palm plantations, and some inhabitants of the region assert that oil palm poisoned the land and that this poison is now, through the snakes, polluting people's bodies. What do these *tramas* say about the effects of war and forced displacement? Do they constitute cultural representations of suffering? Or do they instead express how a collective experience of violence can fundamentally—and lethally—alter the qualities of places and beings that constitute the territory? To address these questions, let us consider the following. Research conducted in Guatemala and Sumatra (Escalón 2014; Ferdman 2014; *Economist* 2019) and various regions of Colombia (Lynch 2015), demonstrates that oil palm plantations augment the densities of snakes well beyond the densities found in natural habitats. There is no doubt that the violent transformation of the forests of Bajo Atrato has provoked irreversible environmental damage. The clearing of huge patches of one of the most biodiverse places in the world, the demise of critically endangered species, declining water quality and quantity, and pollution and soil erosion are only some of the consequences associated with the conversion of natural forests into oil palm plantations. The increased presence of snakes is yet another consequence of war in the region. But I argue that the snakes now teeming in these territories represent more than territorial damage, as this phenomenon exceeds the kind of environmental harm traditionally associated with war. Consider this. People in Curvaradó are well aware of the correlation between the presence of oil palm monocultures and the abundance of snakes. However, when these communities describe the effects of this correlation in terms of an increased number of snakebites now being *tramadas*, it is not simply that they are using their cultural beliefs to arrive at the same conclusion reached by biologists and agronomists—that when palm leaves are cut and gathered into piles they attract lizards, frogs, and their natural predators, so snakes are abundant and they are more likely to bite people. Changes in the landscape have led to significant transformations in the presence of, and relations between, snakes and other animals, which is more than symbolic. Similar to the ideas of fear and estrangement described by Gastón Gordillo (2004) among the Indigenous Toba who worked on sugar cane plantations, *tramas* and the poisoning of lands and bodies are associated with the experience of alienation and deprivation connected with violent capitalist exploitation. And yet *tramas* do not only describe the evils of the paramilitary and capitalist forces that stole and ravaged the lands of communities from the Curvaradó Basin. Instead, *tramas* are evidence of a damage to territory, as there is a lethal poison stealthily destroying both these lands and peoples' bodies. Therefore, the inhabitants' main concern is not only recuperating stewardship over their lands or the control of snakes through the eradication of palm but rather finding methods to properly deal with this poison and heal simultaneously the territory and their own bodies.

Spiritual Masters of Game

On an Indigenous reserve called Eyákera, game animal scarcity has led Indigenous Emberá leaders to request the planting of certain trees as a measure of reparation. According to these leaders, the wrath of the masters that protect game animals was provoked by the constant presence of soldiers in the forest as well as by the logging carried out by invaders sponsored by paramilitary armies, leading these masters to hide game. Through reforestation, the spirits of these trees can help shamans negotiate with the masters that protect animals. As in the case of Curvaradó, the effects of drastic landscape transformations go well beyond the framework of environmental damage. In requesting the planting of trees, people are pointing to a very different kind of harm—one that might appear purely as deforestation and animal disappearance in scientific or conservationist terms but that is in fact far from being limited to territorial damage. It is not that Indigenous communities do not know that upon permitting reforestation, certain plants will attract some insects and pollinating birds, which in turn will permit the presence of species whose fruits will eventually feed large rodents and other potential game animals. For local communities, it would be pointless to let the forest grow if shamans could not then carry out the vital acts of negotiation with the spirits that live there. The state sees reforestation as a measure to increase biodiversity and to restore a damaged landscape. But when Emberá communities suggest planting trees, they intend to provide an environment which will allow for the return of spiritual masters. In this vein, the actions of reparation to the territory must include the recovery of relationships of reciprocity between people and other-than-human beings, and not simply the restoration of a disturbed environmental balance. What is intended as a means to repair the wrongs of war in Emberá territories is not only a recovery of symbiotic relationships between animal and vegetal species but also of relationships of exchange and reciprocity between humans and spiritual masters.

In both aforementioned examples, the affected communities point to a different phenomenon than the one described by environmental scientists. The latter see loss of forests in the case of game animals and overpopulation of snakes in the case of oil palm plantations. Local communities see game scarcity as a breakdown in the relations with spirits, and they see the overpopulation of snakes as the proliferation of animals that embody an evil exacerbated by war. Communities and scientists are describing two different realities that, however, share a partial connection: a profound transformation of local ecologies. The point is that *tramas* and the anger of spiritual masters are symptomatic of a violent shift in the relationship between people, their territories, and the other beings that live there, which is not limited to environmental issues. *Tramas* and the difficulties to negotiate with the protectors of game animals involve existential relationships between different assemblages of beings and are examples of what I described as damage to the territory, whereas the transformation of landscapes—which jeopardizes the sovereignty communities have over their lands—is a characteristic of what can be considered territorial damage. In both cases, it would be simplistic to reduce *tramas* and the anger of spiritual masters to imperfectly formed assumptions of "real" environmental conditions since the current, dire effects of snakebites and the absence of game animals are inseparable from the destruction brought about by war. Therefore, justice and reparation for Indigenous territories should not be limited to the environmental realm, the reestablishment of people's rights, and the protection of their worldviews vis-à-vis nature. Rather, what should be repaired is the set of ties local communities have with the other-than-human beings that form an integral part of their territories and that contribute to the very existence of said territories.

Indigenous Experiences as Law

When two interlocutors speak about what we might take to be the same issue—in this instance, the nature of the damage and the possibilities of reparation—with similar vocabularies but in fact mean radically different things, we face what Blaser describes as an ontological conflict. Ontological disputes occur not because different speakers have different representations of a common world but because "the interlocutors are unaware that different worlds are being enacted (and assumed) by each of them" (Blaser 2009b: 11). This ontological conflict leads to political challenges in terms of the delivery of justice to Indigenous and Afro-Colombian peoples, because when they compel us to consider their territories as victims of war, we are forced to pay attention to damages that extend beyond the human and to the way these peoples' own humanness is constituted through the various emplaced relations they cultivate with other-than-humans. More importantly, through the harm experienced by their territories, Indigenous and Afro-Colombian peoples are addressing concerns about the constitution of the world itself and the nature of the beings that compose it, which is an ontological issue. As such, any attempt from the state to mend the wrongs of war is rendered futile if these wrongs are framed exclusively in terms of cultural representations of the world—that is to say, in epistemological terms. This, as I explain below, is the predicament faced by the multicultural state, and it is one that a political ontological approach can help us understand. The multicultural state aims to be sensitive to the territorial realities harmed by war. But it does so by stabilizing all cultural difference to epistemological difference and reifying nature and culture as two divergent ontological domains.

The inclusion of the territory as a victim challenges dominant modes of politics as it calls into question the radical separation that modern, multicultural states actualize between the realm of nature—with science as its legitimate representative—and the realm of humanity—with secular politics as its principal ally (de la Cadena 2010: 341–42). The language adopted in the victims' law seems also to shift a series of beings—spirits, ancestors, masters of game animals—from the social sphere of particular cosmologies to the political sphere of law, eroding the already porous borders between the living and the material, between subjects and objects, or between what Latour (2004a) conceptualizes as facts and values. Bruce Braun and Sarah Whatmore (2010: ix) argue that the way different life-forms contribute to the world we all share must enter into our current understanding of politics. On the basis of this argument, the inclusion of the territory as a victim becomes a cosmopolitical endeavor: an opportunity to expand the means of responsibility for—or how we respond to (Haraway 2008)—and accountability for—or how we account for (Barad 2011)—the harm provoked to the other-than-human worlds that compose Indigenous and Afro-Colombian territories. I thus interpret the arrival of territory in the realm of law as a seed liable to disrupt the way in which the multicultural state has been regulating the relationships—and building boundaries—between "the real," or the world itself, and cultural representations or worldviews. This seed would be capable of curbing a human-centric approach regarding justice and creating a slightly different awareness about the damage provoked by war and the possibilities of reparation. According to this framework, transitional justice would be based on a less human-centric ethics—that is to say, an ethics not necessarily related to the extension of moral considerations toward other-than-human beings. Rather, it would take as its starting point the human condition of living with and for other nonhumans. This entails reconsidering certain assumed ontological distinctions between people, places, and other beings and centering instead on the mutual constitution of humans, other-than-humans, and places.

In light of this inclusion of other-than-humans in our understanding of war and its aftermath, one ought to start thinking of reparations as a sort of diplomatic endeavor (Stengers 2005b). By diplomacy I mean something different from condescending concessions made by the state in their consideration of the way in which Indigenous and Afro-Colombian peoples think of themselves as actors in the world, or about the kinds of beings they believe can or do exist, or of the actions these peoples believe they themselves and these beings can perform. Instead, diplomacy is a kind of "generative excess" (I. Vargas 2019: 245), an agreement reached by parties with radically different ways of understanding the world, yet who compromise by accepting the absence of a common or absolute ontology as the ground for politics (Stengers 2005b: 194). Diplomacy is a kind of "positive pluralism" (Zournazy 2002: 261), an achievement in which parties are able to entertain their own version of an agreement without undermining their own practices and values. The well-known example of the wasp and the orchid discussed by Deleuze and Guattari (2004) might help us understand this kind of diplomacy. According to these authors, the encounter of an orchid that lures a male wasp by mimicking a female wasp is not just one of imitation—a deceptive flower that has evolved to resemble female versions of a pollinator insect for the propagation of its own species. It is, rather, an instance of becoming: through a mutual capture of code and trace, both entities create a new reality, a rhizomatic association in which the boundaries of one cannot be thoroughly distinguished from the boundaries of the other. Stengers (2010: 29) holds that such an encounter does not create any sort of wasp-orchid unity; rather, "wasps and orchids give each other quite another meaning to the relation that takes place between them." A diplomatic reparation would be exactly that: a generative excess in which each party entertains its own version without contravening the practices and values that sustain their own worlds. As a potential diplomatic event, reparations would allow, on one hand, the state to protect Indigenous and Afro-Colombian territories without framing with the language of rights to cultural difference the damage these peoples situate in other-than-human domains. On the other hand, diplomatic reparations would allow Indigenous and Afro-Colombian peoples to advocate for the protection of these other-than-human beings without necessarily appealing to a language with which modern states are more familiarized (i.e., religious pluralism or environmental sciences). Avoiding a translation that situates at the level of knowledges and concepts (epistemological pluralism) the harm that war provokes in other-than-human worlds would allow the state and Indigenous and Afro-Colombian peoples retain their world while designing reparations that will not cancel their differences (and attain ontological pluralism).

Even before the territory was formally recognized as a victim of armed conflict, other laws made implicit provisions that took into account the special relationship Indigenous and Afro-Colombian communities have with their territories. Consider, for instance, the prevalence that notions such as "spiritual relationships" or "spiritual values" have in the jurisprudence of the Constitutional Court of Colombia or in the Inter-American Court of Human Rights.[4] "Spiritual values" are employed to describe the kinds of relations that extend beyond the material or instrumental significance that Indigenous peoples assign to their lands. In fact, these "spiritual relations" constitute one of the key elements frequently included to legitimize the granting of territorial rights to these peoples. According to Alexandre Surrallés (2017), the concept of spirituality as it is integrated in these types of laws, including the UN human rights system, explicitly recognizes the arrangements of social relations through which Indigenous peoples interact with a wide variety of life, including nonhumans, which are considered in most Amerindian ontologies as beings endowed with some kind of agency and personhood (Descola 2013; Vivieros de Castro 1998). In other words, spirituality in the context of territorial rights is associated with the importance Indigenous communities ascribe to the relations they keep with other-than-human entities for the preservation of harmonic relations with and within their lands. In this way, because the right to territory is usually justified in terms of the spiritual connections Indigenous peoples cultivate with their lands, and because spiritual relations encompass the ties these people cultivate with other-than-humans, it follows that protection of the right to territory is also protection of the existence of these other-than-human entities. Simply put, the notion of spirituality recognizes the existence of a fabric of social relations linking together humans and other-than-humans in a particular locality. Thus, when the spiritual life of Indigenous peoples serves to justify land claims, argues Surrallés (2017: 230), the defense of the right to territory via the defense of Indigenous peoples' rights would include "a nascent extension" of rights to nonhumans.

Some treaties ratified by the Colombian government use the concept of "spirituality" to recognize extended territorial rights to Indigenous peoples, but the consideration of the territory as victim does more. It underscores the lively relationalities of people, places, and other-than-human beings, emphasizing how war hinders the flourishing of wider assemblages of life. Such an approach decolonizes justice because, besides incorporating views that have been historically silenced or marginalized (Izquierdo and Viaene 2018), it calls into question the terms with which modern legal frameworks interpret the damages provoked by war. When the territory is victim, what enters into the realm of justice is not simply respect for the cultural rights of peoples (including the right to use and manage lands) but a set of definitions that go beyond seemingly stable categories such as human/nonhuman or animate/inanimate. This poses what we might call an ontological problem to modern, multicultural states, since for these political entities, beings such as spirits or masters of game animals belong to the sphere of particular worldviews, not to the world itself. In this context, the demands to include Indigenous territories as victims of the armed conflict, along with the possibility of recognizing the damages experienced by other-than-human entities, highlight an inherent limit of multiculturalism and the modern epistemological framework that accompanies it. As a public policy accommodating and regulating cultural diversity, multiculturalism, and modern epistemology are fundamental to the way nation-states divide the public or universal—the interests, practices, and values promoted in the name of scientific and secular knowledge—and the private or particular—the interests, practices, and values attributed to the so-called sociocultural constructions of reality embodied by certain peoples.

Exceeding Multicultural Arrangements

In one of the official editions of the victims' law regarding Indigenous peoples, two high officials from the Colombian government state that the inclusion of the territory as a victim was possible because government and Indigenous authorities held a respectful *diálogo de sabers*, a dialogue among different ways of knowing (Presidencia de la República 2012: 8). For them "this inclusion accounts for the sociopolitical recognition of the diversity of epistemologies and relations between man and Mother Earth that, with their different versions, take precedence in the Indigenous way of thinking" (9). When examining the victims' law from the perspective offered by political ontology—that is, as an instance in which different ontologies strive to maintain their own existence in a political domain (Blaser 2009a, 2009b; Escobar 2015, 2016)—the inclusion of the territory as a victim allows us to imagine various policies of attention and reparation that might result from taking into account a large diversity of other-than-human beings. My take here is that this inclusion represents a political victory for the Indigenous organizations in their quest for justice. However, when damage to territory is wrapped up in concepts such as cosmovision or "the diversity of epistemologies" that the aforementioned state officials referred to, the law adopts a tone in which the multicultural Colombian state simply ensures that they respect the worldviews of particular social actors recognized as legitimate others. The issue is that when those who experience the damage caused by war in terms, let us say, of spirits that have disappeared, snakes that have become more aggressive, or places that are now haunted, these people are not recognized as presenting claims about the very nature of the world and human experience within it (Holbraad 2008) but merely as producing cultural representations about "real" facts. The idea that these experiences correspond to a particular worldview emerges in the epistemological process through which those experiences are translated into modern terms for those who do not inhabit the worlds of spirits or of places endowed with agency.

When invoking, for instance, spirits that have gone mad, communities of Bajo Atrato are calling on different logics of how the universe is constituted (Handelman 2008: 181) and principles "of a being in the world and the orientation of such a being toward the horizons of experience" (Kapferer 2012: 79). In other words, they are making reference to ontic and ontological questions—the constitution of the world and the logic of relationships among the beings that it comprises—and not merely epistemological questions. The turn of the epistemological toward the ontological leads us to consider not so much the way in which certain peoples see war and its aftermath but rather the very nature of damage and the healing of larger communities of life. That is why the dialogue of knowledges that state officials invoke is asymmetrical since the power relations that make this dialogue possible are never called into question. Also, the type of world and reality sanctioned by the modern institutions that make up the nation-state is not admitted as itself subject to negotiation. This may have profound consequences for the policies of reparation to Indigenous peoples.

As an approach, political ontology assumes that conflicts involving dissimilar perspectives have nothing to do with what vision of nature is the closest to reality—that is an epistemological conflict—but rather with the very nature of what exists and can be known—an ontological conflict. In the case of the territory as a victim, if reparation policies are enclosed within a multicultural framework that would just putatively accommodate the plurality of existing cultural representations of the world, the type of damage invoked by Indigenous and Afro-Colombian communities would continue to be confined to the world of their beliefs, meaning that damage to territory would say nothing "real" about the suffering of the territory itself and much less about the impacts on spirits or masters of game animals since none of these entities "really" exist, at least not in the way a modern state assumes. By embracing difference through its modern matrix of knowledge (within which things and events are allocated to either the realm of nature or of culture), the multicultural state circumscribes the possible forms of expression that Indigenous and Afro-Colombian cultural difference may adopt, foreclosing the possibility that this "difference" could challenge both the terms modern institutions mobilize to define reality and the power-laden structures that sustain said version of reality. Put differently, the state recognizes "cultural" differences as long as they do not undermine its own onto-epistemic tenets, bracketing off the question of the world itself from the realm of tolerable difference. In the context of damage to territory and the possibilities of its reparation, this distinction between reality and worldviews limits our ability to effectively address the very definition of damage and its possibilities of reparation. If, within the multicultural paradigm, spirits or places endowed with agency are only allowed to exist as social constructions, can the state deliver justice and reparations that are consistent with the experiences of Indigenous peoples?

From the perspective of cultural politics, the inclusion of the territory as a victim can be perfectly understood in the terms used by the two officials quoted at the beginning of this section: the way in which a multicultural state recognizes the variety of worldviews embodied by Indigenous peoples. From that point of view, one is tempted to recognize the important advances of the national legislation and to interpret this inclusion as an achievement of the Indigenous organizations in the recognition of their rights. From the perspective of political ecology (i.e., the analysis of power struggles in environmental governance), the inclusion of the territory as a victim would mean more. It would provide new ways to understand the disputes over the control, use, and protection of rights of ownership of collective lands and territorial resources, as well as the conflicts that emerge when nature is experienced in radically different ways by different actors. Even from the perspective of political economy, this recognition would problematize the type of hegemonic ideas that have rendered territory and its constituents a collection of natural resources to be exploited. But none of these approaches sufficiently calls into question the nature of the world that can be known since the distinction "between the world (Nature) and its representation (Culture) continues to be affirmed as a universal" (Blaser 2009a: 889). Moreover, cultural politics, political ecology, and political economy frameworks add little to the understanding of the effects of the conflict on territories that communities in Bajo Atrato experience as living entities. Hence the importance of political ontology, since the inclusion of the territory opens possibilities of recognition of entities and effects that go beyond the human, destabilizing the boundaries that state politics have mapped out between the realms of nature and culture.

Conclusion

This article has highlighted the importance of local experiences in the assessment of damage and possible reparations. I have emphasized that Indigenous and Afro-Colombian peoples think about justice in ways that are always embedded in, and in relation to, their own territories, which means that territory becomes somehow a source of legal meaning (I. Vargas 2019) and that the experiences Indigenous and Afro-Colombian peoples derive from it may serve to challenge or *indigenize* (Bacca 2020) some modern imaginaries regarding reparations and justice. I have demonstrated how the indigenization of knowledge arises from situated understandings of territory that in turn shape legal and ethical frameworks regarding the redressing of the damage provoked by war. In other words, different land relations will generate different modes of politics and ethics. Understanding the aftermath of war in terms of the harm caused to the territory and the other-than-humans existing there is fundamental when developing a framework of ethical and political relationships between Indigenous peoples, their territories, and the state, as well as mainstream society. Indeed, the disappearance of the spiritual guardians of forests and rivers says as much about the world itself and how it can be experienced as it does about the values and sensibilities of those who describe the damages of war in these terms.

The recognition of the territory as a victim, as interpreted in my analysis through the lens of political ontology, brings to the fore several different but interconnected aspects of the experiences of war. First, it paves the way for understanding a range of damages that armed conflict has caused in myriad beings; under this framework, the human is merely another agent, and not always the most important. Second, this recognition constitutes an occasion to consider policies of attention and reparation that are consequent with such a diversity of existences. Ontological conflicts produced in such contexts in turn raise questions about what is visible, legitimate, and comprehensible to a state endowed with the power to differentiate between the "real" and the so-called cultural constructions of that reality. Besides creating a certain political awareness, political ontology can be a tool of analysis that, in a scenario of transitional justice, draws attention to the following questions: What types of harms are comprehensible to the modern state, and why are they so? What other kinds of damage caused by the conflict are ignored or sidelined? To what extent might a recognition of those other damages contribute to a new political perspective on truth and reparation? Finally, what ethical concerns would arise from the eventual recognition of a set of entities whose existence goes beyond the human?

While beyond the purview of this article, the recognition of territory as victim further calls into question the way the multicultural state tends to conflate certain experiences regarding territories with particular ethnic identities. By this I mean that the aftermath of war involving and affecting the set of relationships people cultivate with the places they inhabit is not an experience circumscribed to those ethnic groups that the state recognizes as legitimate others (Indigenous and Afro-Colombian peoples in this case). In fact, war has been an experience shared by many rural, peasant societies who are officially excluded from the work of establishing ethnic rights and cultural recognition. The emphasis on political ontology helps us track how certain meanings and experiences arise as a consequence of certain practices (Blaser 2009a), of the cultivation of certain sensitivities and ways of relating with places and with the natural world, all of which is not necessarily confined to specific ethnic identities (Blaser 2014). Last, and although the law does not yet consider it, the possibility of having an experience with a territory that is also victim transcends any of the social conditions that the multicultural state recognizes only to certain ethnic affiliations.

The recognition of the territory as a victim would thus enable actions for the protection not only of cultural frameworks but also of diverse groups of beings and worlds whose existence transcend the traditional legal understanding of cosmology and religion. Through the discourse of victimization and vulnerability, current state reparation policies recognize only a certain type of agency in the territory: one that limits politics to actions of assistance and restitution. However, if we pay attention to the lively relationalities humans and nonhumans undertake in Indigenous and Afro-Colombian territories, politics transcends the social contract within which human relationships are regulated. Rather, Indigenous politics constitute a cosmopolitical endeavor that encompasses agents and beings beyond the human, agents toward whom the multicultural state can hardly act as a legitimate regulator for the plain and simple reason that within the state's regime of truth and power, these other beings—for instance, spiritual masters—cannot even be thought of as real. The harm experienced in and by Indigenous territories compels us to consider a form of justice not only concerned with the restitution of the rights of these peoples but also with the extension to other-than-human beings of the moral and ethical considerations that sustain the very framework of said rights. At the very least, a true cosmopolitical endeavor would make room for a response that takes into consideration the suffering of spirits, agentive places, and other kinds of beings. In this way, policies of reparation to a territory are not reducible to those specific to the human. In the context of a transitional justice framework intended to redress the legacies of past wrongs against Indigenous peoples and their territories, failure to consider more-than-human beings will simply reproduce a hegemonic approach to justice heavily embedded within colonial structures.

Notes

1. Free, prior, and informed consent is an obligation of states that, like Colombia, have ratified the Indigenous and Tribal Peoples Convention, or ILO-Convention 169.
2. Congreso de Colombia, Decreto-Ley 4633, Art. 45. Unless otherwise stated, all translations are my own.
3. Corte Constitucional de Colombia, Sentencia T-380–1993, sec. 12.
4. See, for instance, the sentence T-129–2011 or the case of the Kichwa Indigenous people of Sarayaku v. Ecuador.

References

Bacca, Paulo Ilich. 2020. "Indigenizing International Law and Decolonizing the Anthropocene: Genocide by Ecological Means and Indigenous Nationhood in Contemporary Colombia." *Maguaré* 33, no. 2: 139–69.

Barad, Karen. 2003. "Posthuman Performativity: Towards an Understanding of How Matter Comes to Matter." *Signs* 28, no. 3: 801–31.

Barad, Karen. 2011. "Nature's Queer Performativity." *Qui Parle* 19, no. 2: 121–58.

Blaser, Mario. 2009a. "Political Ontology." *Cultural Studies* 23, no. 5: 873–96.

Blaser, Mario. 2009b. "The Threat of the Yrmo: The Political Ontology of a Sustainable Hunting Program." *American Anthropologist* 111, no. 1: 10–20.

Blaser, Mario. 2013. "Notes towards a Political Ontology of 'Environmental' Conflicts." In *Contested Ecologies: Nature and Knowledge*, edited by Lesley Green, 13–27. Cape Town: HSRC Press.

Blaser, Mario. 2014. "Ontology and Indigeneity: On the Political Ontology of Heterogeneous Assemblages." *Cultural Geographies* 21, no. 1: 49–58.

Blaser, Mario. 2016. "Is Another Cosmopolitics Possible?" *Cultural Anthropology* 31, no. 4: 545–70.

Boyd, David. 2017. *Rights of Nature: A Legal Revolution That Could Save the World*. Toronto: ECW.

Braun, Bruce, and Sarah Whatmore. 2010. "The Stuff of Politics: An Introduction." In *Political Matter: Technoscience, Democracy, and Public Life*, edited by Bruce Braun and Sarah Whatmore, ix–xxxviii. Minneapolis: University of Minnesota Press.

Casey, Edward. 1996. "How to Get from Space to Place in a Fairly Short Stretch of Time." In *Sense of Place*, edited by Steven Feld and Keith Basso, 13–52. Santa Fe, NM: School of American Research Press.

Centro de Memoria Histórica. 2009. *El Despojo de Tierras y Territorios: Aproximación Conceptual*. Bogotá: Comisión Nacional de Reparación y Reconciliación, Instituto de Estudios Políticos y Relaciones Internacionales.

de la Cadena, Marisol. 2010. "Indigenous Cosmopolitics in the Andes: Conceptual Reflections beyond 'Politics.'" *Cultural Anthropology* 25, no. 2: 334–70.

de la Cadena, Marisol. 2015. *Earth Beings: Ecologies of Practice across Andean Worlds*. Durham, NC: Duke University Press.

Delaney, David. 2005. *Territory: A Short Introduction*. Malden, MA: Blackwell.

Deleuze, Gilles, and Félix Guattari. 2004. *A Thousand Plateaus: Capitalism and Schizophrenia*. London: Continuum.

Descola, Philippe. 2013. Beyond Nature and Culture. Chicago: University of Chicago Press.

Economist. 2019. "Palm Oil Is Bad for Biodiversity, with a Notable Exception." March 7. https://www.economist.com/asia/2019/03/07/palm-oil-is-bad-for-biodiversity-with-a-notable-exception.

Escalón, Sebastian. 2014. "Palma Africana: Nuevos Estándares y Viejas Trampas." *Plaza Pública*, January 9. https://www.plazapublica.com.gt/content/palma-africana-nuevos-estandares-y-viejas-trampas.

Escobar, Arturo. 2003. "Displacement, Development, and Modernity in Colombia." *International Social Science Journal*, no. 175: 157–67.

Escobar, Arturo. 2008. *Territories of Difference: Place, Movements, Life,* Redes. Durham, NC: Duke University Press.

Escobar, Arturo. 2015. "Territorios de Diferencia: La Ontología Política de los 'Derechos al Territorio.'" *Cuadernos de Antropología Social*, no. 41: 25–38.

Escobar, Arturo. 2016. "Thinking-Feeling with the Earth: Territorial Struggles and the Ontological Dimension of the Epistemologies of the South." *Revista de antropología Iberoamericana* 11, no. 1: 11–32.

Ferdman, Roberto. 2014. "The Ugly Truth behind Guatemala's Fast-Growing, Super-Efficient Palm Oil Industry." *Quartz*, April 3. https://qz.com/194593/the-ugly-truth-behind-guatemalas-fast-growing-super-efficient-palm-oil-industry/.

Forst, Michael. 2018. "Visit to Colombia, 20 November to 3 December 2018." United Nations Human Rights Special Procedures. https://www.ohchr.org/sites/default/files/Documents/Issues/Defenders/StatementVisitColombia3Dec2018_EN.pdf.

Franco, Vilma, and Juan Restrepo. 2011. "Empresarios Palmeros, Poderes de Facto y Despojo de Tierras en el Bajo Atrato." In *La Economía de los Paramilitares: Redes de Corrupción, Negocios y Política*, edited by Mauricio Romero, 269–410. Bogotá: Corporación Nuevo Arco Iris.

Gordillo, Gastón. 2004. *Landscapes of Devils: Tensions of Place and Memory in the Argentinian Chaco*. Durham, NC: Duke University Press.

Handelman, Don. 2008. "Afterword: Returning to Cosmology—Thoughts on the Positioning of Belief." *Social Analysis* 52, no. 1: 181–95.

Haraway, Donna. 2008. *When Species Meet*. Minneapolis: University of Minnesota Press.

Holbraad, Martin. 2008. "Definitve Evidence, from Cuban Gods." *Journal of the Royal Anthropological Institute*, n.s., 14, no. s1: 93–109.

Ingold, Tim. 2000. *The Perception of the Environment: Essays on Livelihood, Dwelling, and Skill*. London: Routledge.

Izquierdo, Belkis, and Lieselotte Viaene. 2018. "Decolonizing Transitional Justice from Indigenous Territories." *Peace in Progress*, no. 34. https://www.icip.cat/perlapau/en/article/decolonizing-transitional-justice-from-indigenous-territories/.

Kapferer, Bruce. 2012. *Legends of People, Myths of State: Violence, Intolerance, and Political Culture in Sri Lanka and Australia*. Washington, DC: Smithsonian Institution Press.

Latour, Bruno. 2004a. *Politics of Nature: How to Bring the Sciences into Democracy*. Cambridge, MA: Harvard University Press.

Latour, Bruno. 2004b. "Whose Cosmos, Which Cosmopolitics? Comments on the Peace Terms of Ulrich Beck." *Common Knowledge* 10, no. 3: 450–62.

Lefebvre, Henri. 1991. *The Production of Space*. Oxford: Blackwell.

Lynch, John. 2015. "The Role of Plantations of the African Palm (*Elaeis guineensis* Jacq.) in the Conservation of Snakes in Colombia." *Caldasia* 37, no. 1: 169–82.

Ogden, Laura, Billy Hall, and Kimiko Tanita. 2013. "Animals, Plants, People, and Things: A Review of Multispecies Ethnography." *Environment and Society*, no. 4: 5–24.

Presidencia de la República. 2012. *Decreto Ley de Víctimas no. 4633 de 2011*. Colección Cuadernos Legislación y Pueblos Indígenas de Colombia, no. 3. Bogotá: Imprenta Nacional de Colombia.

Proyecto Biopacífico. 1998. *Informe Final General*. 8 vols. Bogotá: Ministerio del Medio Ambiente, Proyecto Biopacífico.

Rangel, Orlando. 2015. "La Biodiversidad de Colombia: Significado y Distribución Regional." *Revista de la Academia Colombiana de Ciencias Exactas, Físicas y Naturales*, no. 151: 176–200.

Sack, Robert. 1983. "Human Territoriality: A Theory." *Annals of the Association of American Geographers* 73, no. 1: 55–74.

Schlosberg, David. 2019. *Sustainable Materialism: Environmental Movements and the Politics of Everyday Life*. Oxford: Oxford University Press.

Stengers, Isabelle. 1996. *La guerre des sciences*. Vol. 1 of *Cosmopolitiques*. Paris: La Découvert.

Stengers, Isabelle. 2005a. "The Cosmopolitical Proposal." In *Making Things Public*, edited by Bruno Latour and Peter Weibel, 994–1003. Cambridge, MA: MIT Press.

Stengers, Isabelle. 2005b. "An Ecology of Practices." *Cultural Studies Review* 11, no. 1: 183–96.

Stengers, Isabelle. 2010. "Including Nonhumans in Political Theory: Opening the Pandora's Box." In *Political Matter: Technoscience, Democracy, and Public Life*, edited by Bruce Braun and Sarah Whatmore, 3–33. Minneapolis: University of Minnesota Press.

Surrallés, Alexandre. 2017. "Human Rights for Nonhumans?" *HAU* 7, no. 3: 211–35.

Todd, Zoe. 2018. "Refracting the State through Human-Fish Relations: Fishing, Indigenous Legal Orders, and Colonialism in North/Western Canada." *Decolonization: Indigeneity, Education, and Society* 7, no. 1: 60–75.

United Nations. 2004. *Informe del Relator Especial sobre la situación de los derechos humanos y las libertades fundamentales de los indígenas, Sr. Rodolfo Stavenhagen, Adición Misión a Colombia*. https://www.acnur.org/fileadmin/Documentos/BDL/2006/4353.pdf.

United Nations. 2010. *La Situación de los Pueblos Indígenas en Colombia: Seguimiento a las Recomendaciones Hechas por el Relator Especial Anterior*. https://www.acnur.org/fileadmin/Documentos/BDL/2010/7377.pdf?view=1.

Valencia, Armando. 2005. "Territorio e Identidad Cultural." *Selva y Río*, no. 2: 15–36.

Vargas, Ivan. 2019. "Forest on Trial: Agency for Transitions into the Ecozoic." In *Liberty and the Ecological Crisis. Freedom on a Finite Planet*, edited by Katie Kish, Christopher Orr and Bruce Jennings, 234–50. London: Routledge.

Vargas, Jennifer. 2016. "Despojo de Tierras Paramilitar en Riosucio, Chocó." In *El Despojo Paramilitar y su Variación: Quiénes, Cómo, Por Qué . . .*, edited by Francisco Gutiérrez and Jennifer Vargas, 121–46. Bogotá: Universidad del Rosario.

Viveiros de Castro, Eduardo. 1998. "Cosmological Deixis and Amerindian Perspectivism." *Journal of the Royal Anthropological Institute* 4, no. 3: 469–88.

Watson, Matthew. 2011. "Cosmopolitics and the Subaltern: Problematizing Latour's Idea of the Commons." *Theory, Culture, and Society* 28, no. 3: 44–79.

Zournazy, Mary. 2002. "Interview with Isabelle Stengers." In *Hope: New Philosophies for Change*, edited by Mary Zournazi, 244–72. London: Routledge.

Daniel Ruiz-Serna is lecturer of anthropology at Dawson College and postdoctoral fellow at the University of British Columbia and Concordia University. He is the recipient of the Governor General of Canada's Gold Medal (2019) and author of *When Forests Run Amok: Violence and Its Afterlives in Indigenous and Afro-Colombian Territories* (2023).

EVOCATIONS of MULTISPECIES JUSTICE

Artwork by Ravi Agarwal and Janet Laurence
Poetry by David G. Brooks

Taralga Road

Early spring, gold
wattle lining the lanes, dams
brimming, fields
emerald-green
and clotted with long-eared lambs, road
deeply pot-holed from the winter rains,
a cyclist
come a cropper
being loaded into an ambulance
seven kilometers north of Kenmore.

"Wombat"
she announces just after Tarlo, and stops, gets out.
I see her in the rearview mirror
turn the body over
before walking quickly back
for gloves and a cloth from the first-aid kit. "She's
dead," she says, "but there's
a tiny paw
reaching from the pouch. I've
got to check." Across the road, two
young black steers, ears
blue-tagged for slaughter,
amble to the fence to watch, then others
and still more, a dozen, twenty, eyes

Cultural Politics, Volume 19, Issue 1, © 2023 Duke University Press
DOI: 10.1215/17432197-10232487

Figure 1 Janet Laurence, *Birdsong* (2006). Assembly of taxidermy bird specimens, suspended acrylic ring. Installation view, Object Gallery, Sydney. Photograph by Keith Saunders.

wide in concern; if they weren't
animals you'd almost think they knew her. Her
baby dead also, she tells me when she returns;
pale
and furless, barely
filling the palm of her hand.

Three kilometers later there's another.
"Their pouches
are so *wet*," she says, then speaks
in awe of the size of their teeth.

By Oberon
there have been two more

Figure 2 Janet Laurence, *Memory of Nature* (2010), detail. Taxidermied owl, acrylic, scientific glass, tulle, wood, oil paint. Art Gallery of New South Wales collection.

and almost a dozen roos. Some
we stop at, others you can just tell
it's far too late; or there's
a truck on your tail, or the road's so narrow
there's no space to pull over,
let alone any place to run.

Near the turnoff to Jenolan, dusk coming on,
we stop at a young swamp wallaby, head
crushed by a curb-side wheel, the road
a single lane, then, our
examination done, step
back to allow a four-wheel drive to pass, watch
as it grinds her—eyes, ears, brain—
even deeper into the gravel. Her joey
who'd been
still breathing
dies in our hands.

Figures 3 and 4 Janet Laurence, *Fabled*, from the After Eden series (2011). Altered camera trap images, ink on archival paper. From the Flora and Fauna International Residency, Acheh, Sumatra, Indonesia.

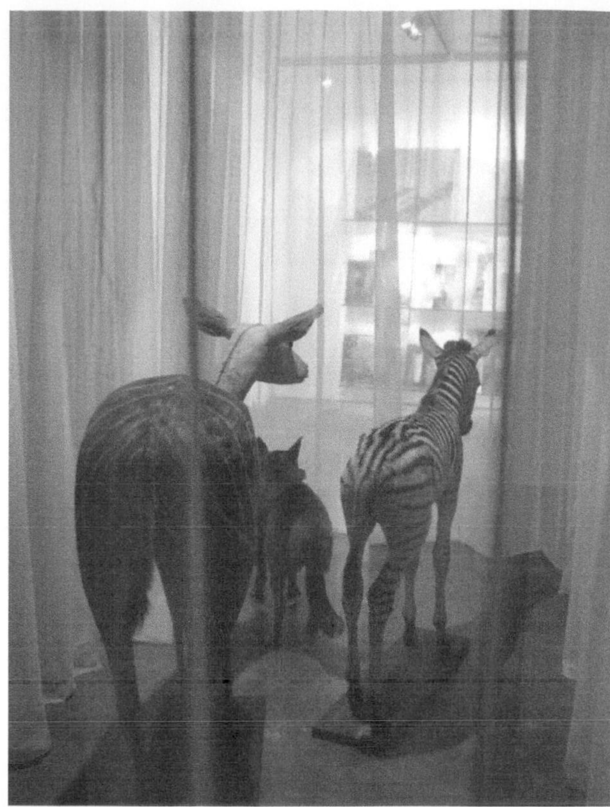

Figure 5 Janet Laurence, *Fugitive in Light* (2002). Taxidermy specimens borrowed from the South Australia Museum. Installation view, Eden and the Apple of Sodom exhibition, University of South Australia Museum, Adelaide.

Figure 6 Janet Laurence, detail from After Eden exhibition (2011). Altered image from John Gollings original mine photograph.

Figure 7 Ravi Agarwal, *Ambient Seas Diary*, Else All Will Be Still series (2015). Photo collage.

Figure 8 Ravi Agarwal, *Landfill I*, Trace City series (2017). Photographic inkjet print.

Figure 9 Ravi Agarwal, *Na'dar Landscapes Series—1*, Na'dar/Prakriti series (2018). Lithograph.

Figure 10 Ravi Agarwal, *Na'dar Landscapes Series—7*, Na'dar/Prakriti series (2018). Lithograph.

Figure 11 Ravi Agarwal, *Salt Pan*, Else All Will Be Still series (2015). Photographic inkjet print.

Figure 12 Ravi Agarwal, *Power Nature I*, Na'dar/Prakriti series (2018). Photographic inkjet print.

Ravi Agarwal is an interdisciplinary artist, environmental campaigner, writer, and curator. His works use photography, video, text, and installation and have been shown widely, including at the biennials of Havana (2019), Yinchuan (2018), Kochi (2016), and Sharjah (2013), along with dOCUMENTA XI (2002). He has curated large public art projects, such as the *Yamuna-Elbe* twin city project (2011) and *Embrace Our Rivers* (2018), an Indo-European project, and was the photography curator for the Serendipity Arts Festival in 2018 and 2019. He recently curated *New Natures: A Terrible Beauty Is Born* (2022) at the Goethe Institute and CSMVS Museum, Mumbai, and *Imagined Documents* (2022) at Les Recontres d'Arles, France. He has edited the collections *The Crisis of Climate Change* (2021) and *Embrace Our Rivers* (2017) and special issues of journals, including "Art and Ecology" (*Marg*, April 2020) and *IIC journal* (Spring 2020) and has been published in *The Routledge Companion to Contemporary Art, Visual Culture, and Climate Change* (2021). Agarwal is the founder-director of the environmental research-advocacy group Toxics Link (https://www.toxicslink.org) and recipient of the UN-IFCS Award for Chemical Safety and the Ashoka Fellowship.

Exploring notions of art, science, imagination, memory, and loss, ***Janet Laurence***'s practice examines our relationship to the natural world through site-specific, gallery, and museum works. Working in varying mediums, Laurence creates immersive environments that navigate the interconnections between diverse life-worlds. Her work explores what it might mean to heal the natural environment, albeit metaphorically, while fusing a sense of communal loss within this search. Laurence has been a recipient of Rockefeller, Churchill, and Australia Council fellowships and the Alumni Award for Arts, University of New South Wales. She was the Australian representative for the COP21/FIAC, Artists 4 Paris Climate 2015 Exhibition, during which she exhibited a major work—*Deep Breathing: Resuscitation for the Reef*—at the Muséum national d'historie naturelle in Paris. In 2019 she had a major solo survey exhibition at the Museum of Contemporary Art Australia, *After Nature*.

David G. Brooks, an Australian poet, short fiction writer, essayist, novelist, and advocate for nonhuman animals, taught for many years at the University of Sydney, where he is now an honorary associate professor. His recent works include *Open House* (poetry, 2016), *The Grass Library* (memoir/animal rights, 2020), *Animal Dreams* (selected essays, 2021) and *Turin: Approaching Animals* (meditations, 2022). He lives with rescued sheep in the Blue Mountains of New South Wales. His website is found at https://davidbrooks.net.au.

POLITICAL PLANTS

Art, Design, and Plant Sentience

Sria Chatterjee

Abstract **This essay considers a series of examples of contemporary and early twentieth-century artistic projects done in collaboration and conversation with plant scientists around the theme of plant sentience. In particular, it zooms in on the work of the Indian biophysicist Jadagish Chandra Bose and the Indian artist Gaganendranath Tagore in the 1920s and the Italian plant scientist Stephano Mancuso and German artist Carsten Höller in the 2020s. The essay has four interconnected aims. The first is to investigate how and why plant sentience is visually and spatially represented by artists. The second is to show through two broad examples how plant science can be and has been co-opted to serve different political, economic, and ideological positions. The third and broader aim of this essay is to counter a widespread ethical assertion in environmental humanities and animal studies that destabilizing human-nonhuman binaries intrinsically lends itself to projects of environmental justice by encouraging humans to coexist more equitably with other species. In other words, we should not assume that artistic production is spontaneously aligned to ethics of multispecies justice. The fourth and concluding aim is to make the related argument that plant sentience and other ways of knowing and relating across species need to be understood within the context of colonial and extractive histories.**

Keywords **art and design, plant sentience, biomimicry, politics, colonialism, nationalism**

Part 1
Introduction

Plant sentience has long divided both scientific and lay circles. Following a resurgence of interest in plant sentience with bestselling books such as Daniel Chamovitz's *What a*

Cultural Politics, Volume 19, Issue 1, © 2023 Duke University Press
DOI: 10.1215/17432197-10232502

Plant Knows (2012), Peter Wohlleben's *The Hidden Life of Trees: What They Feel, How They Communicate—Discoveries from a Secret World* (2016), and others, plant sentience also entered the popular as well and artistic imagination with some force. Many of these artworks, some of which I discuss in this essay, foreground plants as actors in their own right in the artwork rather than purely as representations produced by the artist. Despite long-held skepticism around questions of whether plants are conscious beings, plant life and biopolitics have garnered a small yet potent body of scholarship that can be identified as the plant humanities. In recent years, a number of philosophers and literary theorists working within the Western philosophical canon have argued for plants as agential beings: Michael Marder offers a long sweep through classical theory in *Plant-Thinking* (2012), Matthew Hall explores a political theory turn in *Plants as Persons* (2011), in *Life of Plants* Emanuele Coccia (2018) argues that plants are a vital part of metaphysical life, and Elaine Miller provides a feminist account of German Romanticism in *Vegetative Soul: From Philosophy of Nature to Subjectivity in the Feminine* (2002). Framing plants as agential beings capable of meaning-making, these works have broadly advocated for working toward a phytocentric rather than an anthropocentric position. They have sought both to recast the bases on which moral consideration, for example, is afforded and to demonstrate that plants in fact possess many of the affordances that were said to ground moral considerability. For example, whereas the British philosopher Jeremy Bentham's insistence that it was the capacity to feel and not to reason that justified moral considerability has been the basis for the animal turn in ethics, plants remained excluded.

As valuable as this work has been, critics have noted that works by Marder and others took the canon to be the Western philosophical canon—one that considers neither Indigenous nonanthropocentric thought nor the diversity of plant life (see Hamilton 2016; Pettman 2013). While anthropologists and scholars working in Indigenous studies have long tried to decenter anthropocentric discourses (Viveiros de Castro 2012; Kohn 2013; Kimmerer 2013; Todd 2015), it is worth saying at the outset that the majority of scholarly and artistic engagement with plant sentience has remained largely Eurocentric. The expanded field of plant humanities includes an engagement with plants as a part of a more activist approach to the environment and food studies such as Marie-Monique Robin's *The World According to Monsanto* (2008), Wenonah Hauter's *Foodopoly: The Battle over the Future of Food* (2014), and Frederick Kaufmann's *Bet the Farm: How Food Stopped Being Food* (2012). Scholars such as Jeffrey Nealon attempt to take plants to be a linchpin for thinking about biopolitics. Nealon (2016: 121) argues that "contemporary neoliberal nation-states are designed to protect and serve the very privatizing biopolitical pathos of nineteenth-century subjectivity, focused on the life of the individual organism." However, the question of plant sentience itself—the fact that plants can feel and make decisions—has been considered broadly apolitical. This essay works against the two assumptions outlined above to show that plant sentience is neither a product solely of the Eurocentric imagination, nor is it apolitical.

The broad aim of this essay is twofold. First, through a series of contemporary examples, I investigate how plant sentience is visually and spatially represented by artists, often in collaboration or conversation with plant scientists. To what end, I ask, are these artistic interventions aimed? Looking closely at artistic projects by Italian plant physiologist Stefano Mancuso in collaboration with German artist Carsten Höller, and others, I trace some of the ways in which artistic projects around plant sentience aim to make human beings more aware of the agency of plants but also act as a stepping stone for use in the fields of design and innovation such as in the manufacture of plant robots. This leads me to the second crux on which the essay rests, and that is the argument that plant science can be and has been co-opted to serve different political, economic, and ideological positions. Exploring instances from the contemporary moment and going back to the early twentieth century to explore the work of Indian artist Gaganendranath Tagore and biophysicist Jagadish Chandra Bose, I explore how plant sentience has been co-opted or mobilized in different ways—by causes as diverse as bioengineering to anticolonial Hindu nationalism. Mancuso, in the catalog that accompanies the *Florence Experiment* exhibition, quotes the work of Jagadish Chandra Bose, reproducing some of Bose's scientific illustrations published in 1922 to "illustrate nervous excitation in plants" (Galansino 2018). To trace the genealogy of what he and a group of researchers in 2006 named "plant neurobiology," Mancuso goes back to Charles and Francis Darwin's *The Power of Movement in Plants* (1880) and to Jagadish Chandra Bose's (1913) research into plant sentience or irritability.[1] In the early twentieth century, Bose conducted conclusive experiments to prove that plants respond to external stimuli. Bose, who pioneered the investigation of radio waves, is perhaps even better known for extending his work on the physics of electromagnetic radiation to experiments on the life-processes of plants.[2] Building on the genealogical link that Mancuso already makes to Bose, the two parts of this essay focus on case studies about a hundred years apart that open up Mancuso's and Bose's work in collaboration and conversation with artists. The reason for choosing the two is not to simply compare them as equivalent examples at different points in time and geographical location but to show in two very different instances how scientific research into plant sentience is translated and communicated to particular audiences through artistic endeavors and also how it is then co-opted toward different agendas in the cultural, political, and economic sphere. Broadly, this essay counters a widespread ethical assertion in environmental humanities and animal studies that destabilizing human-nonhuman binaries intrinsically lends itself to projects of environmental justice by encouraging humans to coexist more equitably with other species. Ultimately, I argue that plant sentience and other ways of knowing and relating across species need to be understood within the context of colonial and extractive histories.

Donna Haraway (2008: 244) tells us that "if we appreciate the foolishness of human exceptionalism then we know that becoming is always becoming with, in a contact zone where the outcome, where who is in the world, is at stake." Other scholars argue that in accepting multispecies agency, we need to recognize the complex relationships between multiple organisms and understand "the human as emergent through these relations" (Ogden, Hall, and Tanita 2013: 6). Anthropologist Natasha Myers (2017: 300) names the Planthroposcene as "a call to change the terms of encounter, to make allies with these green beings" that has "the potential to stage both new scenes of, and new ways to see (and even seed) plant/people involutions." While in agreement with these assertions, this essay works against the implicit assumption in much of the multispecies literature that once we recognize the sentience of other beings, we should experience them as subjects of moral considerability and even kin. Rather, the essay demonstrates that the human recognition of plants as sentient, agential beings does not automatically lead to them being considered within a larger framework of ethics and justice. Even as scientists make more discoveries about the complexity of plants, it is in the arts that we see what seems to be the effort to do the cultural transformative work of shifting how humans understand themselves and others. We should, however, not assume that artistic production is automatically aligned with ethics of justice. As will be demonstrated in this essay, in fact, it can also be inveigled into logics and systems that exacerbate injustices.

Rather than providing a clear path toward multispecies justice, the essay shows how muddy the waters are by exposing some of the results that engagement with plant sentience has produced. It makes evident the gap between the representation of plants as agential beings in cultural discourse and the representation of plants in the ethical, political, and legal realms. It enacts a middle step in the ostensible continuum between plant agency and plant justice by raising questions around the ethics of co-option. The processes of co-option I describe are often well intentioned in wanting to champion plant sentience and agency. However, they are always mediated by a range of existing and essentially human-centered discourses and relationships with colonial, nationalist, and capitalist world orders.

Plants as Agents

The vital functions of plants, including sentience and decision-making, remain invisible and imperceptible to humans because they happen on a scale and speed that lies outside the most immediate modes of human sensing. The question of how to represent plant sentience, therefore, has been one that has recently of considerable interest to contemporary artists working within the plant humanities. To dispel the notion that plants are either mute or silent, for instance, various artists have rigged up simple electrical circuits to make the hum of plant life audible.[3] As early as in 1902, Bose (to whom this essay will return in more detail), in his treatise *Responses in the Living and the Non-living*, had written eloquently about plant listening. He argued that plants grew more quickly when exposed to pleasant music and gentle whispers, and poorly when exposed to harsh music and loud speech.

Since the 1960s and 1970s, the use of biosensing electronic devices on plants has been a tried and tested way to create connected circuits that interface between technological systems, human sensory capacities, and plant life (Kahn 2013: 240). *Sonic Succulents: Plant Sounds and Vibrations*, an installation inspired by the Sydney-based plant scientist Monica Gagliano and created by Adrienne Adar at the Brooklyn Botanical Gardens in 2019, allows visitors to hear plant sounds through headphones when they touch the plants. Adar affixed handmade sensors to a selection of plants such as the ponytail plant and the golden barrel cactus and objects within their environment, and then amplified them through audio equipment. Visitors were invited to put on headphones, and when they touched a plant, they were able to hear the plant interacting with them. Other parts of the exhibit also allowed the visitor to go beyond plant–human interaction to plants' "natural botanic rhythms." For instance, in one section, visitors were invited to tune into the sounds of a giant yucca plant growing. In another, they could listen to stalks of corn and plants absorbing water via their roots. For Adar, the premise of the installation was that by interacting with audible plants, visitors would have access to a new perception of these photosynthesizing organisms, not simply as inanimate mute objects for humans to control but as living coinhabitants. Installations such as Adar's and others create an experiential space in which a human visitor is made to interact with plants in a way that goes beyond treating plants as "backgrounded" (Plumwood 2002) into ornamental sceneries (Wandersee and Schussler 2001).

In the same way that Adar tries to render the inaudible world of plants audible, Chiara Esposito's *The Dream of Flying* (featured in the 2013 Ars Electronica Festival) attempted to make the slow and typically imperceptible responses to stimuli visible. Using a digital interface, Esposito enabled an air plant (*Tillandsia* spp.) and a common dandelion (*Taraxacum* spp.) to control their own movements through a body extension: a small flying device. The interface measured the electrical activity of the plant and used the gathered data to control the flight. Installations such as Esposito's and Adar's allow plants to work as both agent and sensor, creating a common space in which visitors and plants interact at a scale that is perceptible to human senses. An increasing number of art installations and artistic experiments have become more specifically invested in questions of plant sentience, especially research into plant cognition, acoustics, learning, and memory (e.g., Gagliano 2013; Gagliano et al. 2014; Trewavas 2006).

In 2006, Stefano Mancuso and five others chose to coin the term *plant neurobiology* to define a new field of inquiry arguing that plants do indeed possess sophisticated behaviors that cannot be explained away by genetic and biochemical mechanisms (Brenner et al. 2006). These researchers insisted that the electrical and chemical signaling systems identified in plants are in fact homologous to those found in the nervous systems of animals. This decision received vehement pushback from other prominent plant scientists who, disregarding the call of homology, argued that plant neurobiology was a misnomer because plants did not possess structures such as neurons, synapses, or a brain (Alpi et al. 2007).[4] The conclusions that emerge from experiments in plant sentience from as early as the 1900s indicate that not only do plants feel, they also display agency in responding differently to different kinds of stimuli. In collaboration with the artist Carsten Höller, Stefano Mancuso worked on a large-scale art-science installation called *The Florence Experiment* (2018). Mancuso has worked on other such exhibitions to produce the conditions in which plants and humans share not only the same space but also, to some extent, the same information and experience. Mancuso's collaborations with artists, designers, and innovators, which I discuss in some detail in this paper are interesting because it is in Mancuso's attempts to communicate to a wide audience through visual and spatial installations how plants "think" (i.e., his artistic collaborations) that eventually leads to further collaborations in the more applied fields of biotechnology.

The Florence Experiment

In *The Florence Experiment*, Mancuso and Höller try to create conditions in which visitors can, although briefly, be aware of what it feels like to be in a symbiotic relationship with a plant, and they try to make clear to people that plants are extremely sensitive organisms. The exhibition opened at the contemporary art museum, Palazzo Strozzi in Florence, Italy, in 2018. In his work Stephano Mancuso cites and references Jagadish Chandra Bose, who this essay briefly introduced earlier and will come back to in greater detail in part 2 of this article. Bose, who first used the term *plant nerve* (locating the nervous route for long-distance electrical signaling in the vascular tissue, specifically the phloem), in some ways forms a basis for *The Florence Experiment* and other installations Mancuso has collaborated on at the nexus of art and plant sentience. An avid advocate of plant neurobiology (Brenner et al. 2006), Mancuso is the founder and director of the International Laboratory of Plant Neurobiology, headquartered in Florence.

The Florence Experiment consisted of two parts. In the first part, two steel and polycarbonate slides were made to snake down from the palazzo's third-floor balcony to the ground floor. Visitors to the exhibition or experiment who wanted to participate were invited to step into the slide from the third floor and were given a bean plant to carry as they whizzed down the slide. Once they were out, they would hand the plant over to a set of scientists waiting in the "Plant Delivery Room." At the same time, some plants were sent down on their own. The scientists then analyzed the plants by looking at several physiological parameters, from growth to photosynthesis and emission of messages in the form of volatile molecules, which may or may not have been affected by the emotions experienced by the attendant visitor on the ride, comparing them with plants that were unaccompanied.[5]

The second part of *The Florence Experiment* consisted of two identical cinemas installed in a different part of the palazzo. One cinema showed scenes from scary horror films and the other from lighthearted and happy comedies. Visitors were invited into the cinemas, and from each cinema a small part of the volatiles (molecules traveling in the air) produced by the viewers were collected and tested separately for their composition in the on-site lab. The rest of the air containing volatiles from the two different cinemas was sucked out through aspirator hooks and carried through different ducts to the palazzo's facade. Here, wisteria plants in big pots were set up to climb a large steel wire which bifurcated at a Y. At the bifurcation, the plant had to decide whether to turn right or left according to the volatiles, with those from one cinema going to the right side of the fork and those from the other going to the left. The aim, as Mancuso said, was to see if the volatiles produced by humans under different psychological conditions are able to influence the behavior of plants. Both parts of *The Florence Experiment* are based on the idea that plants somehow respond to the emotions of human beings. In not claiming to be a controlled scientific experiment, *The Florence Experiment* was "exploring" elements of a scientific experiment, with more attention to communicating this to visitors rather than obtaining a scientific result. In interviews, Höller reiterates that *The Florence Experiment* is an art show, not a scientific study. "It is a proposition to rethink our understanding of things" (Rysman 2018). What does this rethinking *do* for plants?

In Mancuso and Höller's "art show," Mancuso wants to build a "vegetal approach." He wants visitors to see the symbiotic alliance-forming capacities of plants as a strategy for future living. The "vegetal approach" echoes Donna

Haraway's (2003: 7) notion of working toward a "barely possible but absolutely necessary joint futures," of learning to live well with others. However, I argue that for Mancuso and Höller, rethinking plant intelligence is not about the creation of a new political imaginary. Even though *The Florence Experiment* allows humans to see plants as agential beings, the buck stops there. Growing out of a long engagement with Western ontologies, Mancuso's approach furthers the paradigm in which plant sentience and plant intelligence in fact become a gateway to an interventionist redesign of plants to further a human-centric revival of a greener planet. While contemporary art installations may make publics more aware of plants as agential creatures capable of decision-making, design-centered innovation, which I turn to next, aims to *apply* this knowledge to a variety of different ends. Specifically, I place my discussion of these artistic interventions in conversation with questions around biomimicry and biomimesis, and with some of the critical scholarship around these ideas.

Designing Plants

The Italian bioengineer Barbara Mazzolai, in partnership with Mancuso and a group of researchers at the Italian Institute of Technology, have been working on the "Plantoid project" since 2012 (Mazzolai and Mancuso 2013). Mazzolai is interested in developing bio-inspired robots that can explore the soil and the air, cover large areas, detect pollutants, and predict their impact on human health and environment. Research on plant intelligence allowed Mazzolai and her team to introduce a new paradigm to the field of robotics, which they term "moving by growing" (Mazzolai 2019; Mazzolai and Laschi 2020). This draws from the fact that the movement of plants is based on the addition of material, which results in plant growth. In the same way, the plantoid has roots that grow from their tips by way of a miniaturized 3D printer that changes the viscosity of PLA filaments by increasing temperature. The tips of the roots have incorporated sensors, a motor that pulls the filaments, gears to shift their position, and a small resistor that increases the temperature in order to change the filaments' viscosity. Their aim is that plantoid robots can be used in a variety of different ways—from environmental monitoring to space exploration, and perhaps even medical applications (Mazzolai and Mancuso 2013). In short, by putting plant intelligence to work through an integration of sensors in robots, a different techno-natural species is created.

Artists and theorists working in plant humanities explore the philosophical implications of plant intelligence and decision-making as "distributed," in a similar way to "swarm" intelligence, which characterizes the behavior of social insects such as termites and bees to foreground the primacy of "the connections between the individual workers that form a network and changes in communication [that alter] the behavior of the whole colony" (Trewavas 2006: 7). Does memory require a brain, or can we define it simply as an act of remembering that can be triggered not simply by an animal brain but by different processes that produce similar results (Gagliano et al. 2014)? Feminist science studies scholars such as Katherine Hayles (2002: 4) argue that a "distributed" concept of intelligence also recognizes that cognition is not constrained by the processes of the brain alone but rather is a "systemic activity, distributed throughout the environments in which humans move and work."

For Mancuso, despite seeming to be an oxymoron at first sight, plant intelligence

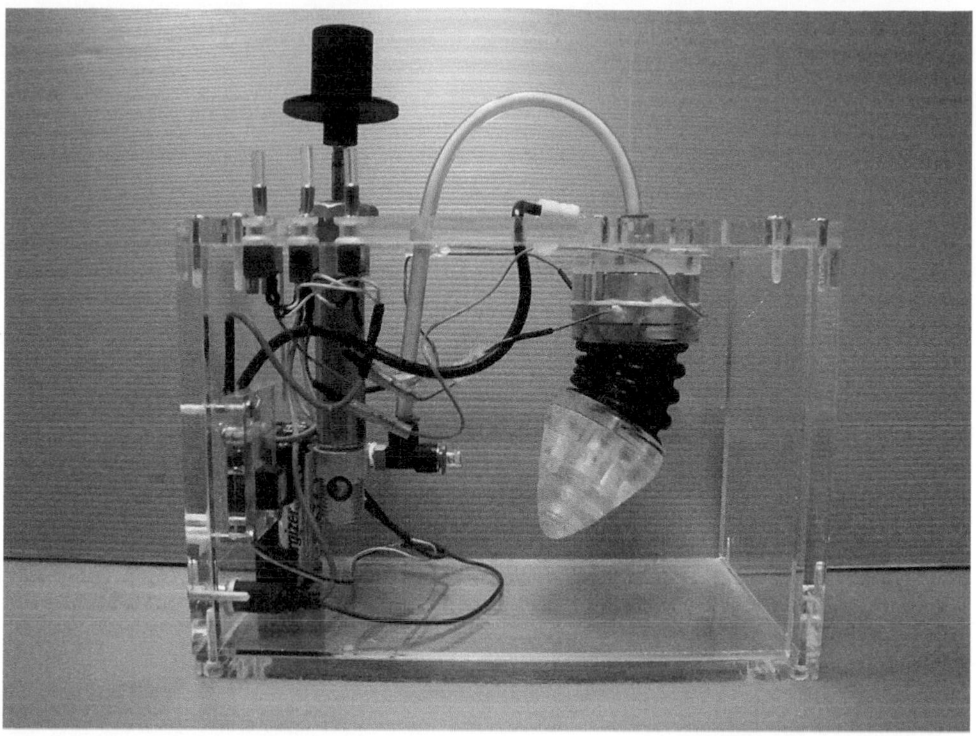

Figure 1 Plantoid. © Plantoid Project 2012. Supported by the Digital Office of the Istituto Italiano di Tecnologia.

is the plant's ability to solve problems. In the place of a brain, what Mancuso is looking for is a sort of distributed intelligence, as we see in the swarming of birds or colonies of insects (Galasino 2018). Similarly, Michael Marder (2012: 154) rejects the idea of a "formal plant-intelligence," and instead imagines the "non-conscious life of plants" as a kind of "thinking before thinking." According to evolutionary ecologist Monica Gagliano, even though plants lack brains and neural tissues, they possess a sophisticated calcium-based signaling network in their cells. This network is similar to animals' memory processes, and therefore, when she claims that plants can remember, the onus is on us to reconsider how we define memory.

By focusing on the otherness of plants rather than on their likeness when it comes to distributed intelligence, Mancuso claims that we stand to learn valuable things and develop important new technologies. He uses the term *bio-inspiration* to talk about the application of plant intelligence to design, in this case of technologies and systems that are "networked, decentralized, modular, reiterated, redundant and green" (Pollan 2013: 104–5). Interfacing between vegetal life and technology, plant ontology serves both as inspiration for newer models of biotechnology and is also put to work as nodes within an ultimately extractive economy.

Mancuso's collaborator, Barbara Mazzolai is also the coordinator of *GrowBot*, a robotics project inspired by the moving-by-growing abilities of climbing plants. It aims to develop low-mass and low-volume robots that are capable of anchoring themselves, negotiating voids, and more generally climbing, where current climbing robots with legs, wheels, or rails would get stuck or fall. Bio-inspired and biomimetic soft machines are developed on functions and working principles that have been abstracted from biological organisms that have been evolving for more than 3.5 billion years. The goal is to create robots that are able to not only sense and respond to the environment but also autonomously and continuously adapt in response to environmental stimuli (see Sadeghi, Mondini, and Mazzolai 2017; Must, Sinibaldi, and Mazzolai 2019; Mazzolai et al. 2020). Mancuso's focus on the ability of plants to solve problems is, however, less kin-making than soft-capitalizing, whereby experiments in plant science and design feeds into a larger extractive innovation industry.

Biomimicry and Vegetal Labor

Human geographer Stephanie Wakefield warns against considering the blurred boundaries between humans and nonhumans in the urban environment as inherently "good" (see Wakefield in Johnson 2014). She emphasizes the growing prevalence of "resilience" thinking from within neoliberal paradigms and argues that it is important not to pay attention to how more-than-human assemblages are taken up and registered through dispositifs of power and technoscience. "Dispositifs," Johnson and colleagues (2014) write, "work not by imposing order on a preexisting field but by arranging and producing the terrain itself, through a double biopolitical movement that simultaneously posits a vision of life and constructs it." Biomimicry destabilizes and decenters the human by learning from and valorizing natural processes. However, the ways in which biomimetic practices are co-opted into capitalist and military-industrial frameworks constitute the more urgent concern. Mazzolai's group advocates for biomimesis, the techno-scientific framework that "reverse-engineers" plant life, instrumentalizing knowledge about plant intelligence. Biomimicry, in particular, has the potential to produce a whole new category of workers. The Biomimicry Guild, set up by Janine Benyus, who was awarded the UN's "Champion of the Earth" award in science and innovation and who genuinely seems to believe that biomimesis is the answer to the historical rupture between humans and their habitat (Benyus 1997), is a consulting firm providing services to enterprises and industries, offering possibilities for "innovation and sustainability." While the consultants at the guild—or nature's translators, as they call themselves—promise to "unlock" knowledge developed in the biological sciences to develop designs and strategies to fit every corporate need and charge exorbitant sums of money for their services, what of their nonhuman collaborators?

In the same way that microbes and viruses are agential beings, biotechnological hybrids are also bioengineers.[6] This opens up various questions. For example, who regulates the actions of bio-technological hybrids, their living, dying, and multiplying? Does flourishing with bio-robots and green computing put other species at risk? What does it mean for plants as organisms if they can in fact be surpassed by more useful plantoids? Advocates of biomimesis insist that it is the answer to a less anthropocentric world. Harnessing theoretical writing in posthumanism and animal studies, destabilizing or even erasing the categories of the human and animal, and recognizing the shared limits and vulnerabilities of both categories (Agamben 2004; Oliver 2007; Wolfe 2010), biomimesis enthusiasts claim that biomimicry has the potential to radically transform human subjectivity and to work collaboratively with nonhumans for a more democratic and sustainable future (Benyus 1997). Yet what type of collaboration is this? If biomimesis was indeed to become a dominant techno-natural fix for the future, what we need is to develop a framework of politics and ethics that is adequate to these new forms of production.

Harold Perkins in a 2007 *Geoforum* article, discusses how Marx holds labor to constitute the common ground of all living beings, who transform their environments by the simple act of extracting matter to sustain their basic metabolic needs. Perkins urges us to take seriously Marx's suggestion and examine its implications for understanding the power relations that unfold through the articulation of different types of labors—whether deliberate or instinctive (Perkins 2007; Ernwein, Ginn, and Palmer 2021). As Elizabeth R. Johnson (2010) argues,

> While the narrative of ecological salvation remains very much in play in these contexts, advocates of the bio-inspired enhancement of industrial production has most often emphasized sustaining or expanding capitalism over sustaining or better serving human and nonhuman life-as-such. Beneath the rhetoric, advocating for green capital seems less about transforming our modes of production than intensifying them.

Projects such as Mazzolai's feed into a recent turn toward postnatural environmentalism. This turn (rightly) dismisses the nature-culture divide but then uses this premise to promote a type of environmentalism in which the natural world is made dependent on human interference and innovation (Nordhaus and Shellenberg 2011; Ellis 2013, Kareiva, Marvier, and Lalasz 2012). This kind of postnature (wrongly, in my view) assumes a level playing field in which the world is made up of heterogeneous living and inert entities. Rather, as it plays out here, the postnatural orientation toward a future, entangled common world is integrated into the logic of the existing neoliberal order. The rhetoric erases inequalities (between humans and other sentient beings) and colonial histories of power and domination, even as they persist into our current geopolitical and economic structures. Considering "Nature" a colonial formation, Rosemary Collard, Jessica Dempsey, and Juanita Sundberg (2014) argue that recognizing entanglement is not enough. They call instead for pluriversal alliances (Blaser, de la Cadena, and Escobar 2013) that benefit the widest range of beings, in place of putatively universal approaches that in fact sustain a human-oriented extractive economy. Recent work in more-than-human political economy has attempted to rethink nonhuman agencies not just as a form of "lively capital" (Haraway 2008) usable as a resource but as coconstitutive of the broader economic realm "from the outset" (Barua 2019: 651). While the fact that agricultural production and the manufacture of biofuels are led not just by human labor but by the active participation and manipulation of tissues, cells, and other biological materials themselves (Rajan 2006), the potential of adapting plant sentience toward new forms of production and labor adds a new category of vegetal work within broader bio-economies. Artistic interventions such as *The Florence Experiment* and its cognate bioengineering projects urge us to revisit Braun's (2009: 31) caution that "there is no hard and fast rule that a particular ontology leads necessarily to a particular politics, but neither can any ontology be said to be neutral." If artists and scholars are working toward engaging broader publics in relationist ontology, I argue that it is also important to consider what the recognition of plant sentience by humans can contribute to a framework of more-than-human multispecies justice.

Biomimicry is featured as a key method for generating a more ecologically sustainable economy.[7] Often cited examples of this include butterfly wing–inspired computer screens that reflect light and save energy, spider-inspired materials that transform the manufacture of durable fabric, and termite mound–inspired building designs that shrink heating and cooling costs (Goldstein and Johnson 2015). Some designers go further and claim that they are shifting from prioritizing human needs to exploring the network of relations that exists in a multispecies ecology (Forlano 2016). This brings home the nuanced terrain we inhabit when it comes to questions of communicating and adapting plant sentience. Although plant-based climate solutions, green infrastructure, and ecosystem services crowd our current landscape, plant sentience has been co-opted by a range of individual and groups ever since its early inception. Part 2 highlights the longer histories of the way plant sentience has been co-opted. In particular, it focuses on the political and religious facets of this history, which usually get left out of conversations and scholarship on plant sentience and more-than-human approaches more broadly.

Part 2

J. C. Bose and Gagenendranath Tagore: Inanimate Scream

Going back about a hundred years to the early 1920s, when plant sentience was a new and contested field of study, part 2 of my essay considers how in a very different context, ideas around plant sentience seeped into the artistic and cultural fields and were also co-opted, but to very different ends. In 1926, Jagadish Chandra Bose published *The Nervous Mechanism of Plants* (dedicated to the poet and educator Rabindranath Tagore), in which he pushed his conclusions one step further. He maintained that plants had "a system of nerves that constituted a single organized whole" (Bose 1926: 121). Arguing that plants have what he called a well-developed nervous system, Bose claimed that (a) all plants coordinate their movements and responses to the environment through electrical signaling, and (b) all plants are sensate, neuronal organisms that actively explore the world and respond to it through "a fundamental, pulsatile, motif in which oscillations of electric potential are coupled with oscillations of turgor pressure, contractility, and growth" (Shepherd 2009: 133). Hackles were and continue to be raised in scientific circles by the use of the word *nervous*, as with words such as *intelligence*, *sentience*, and *consciousness* in relation to plants.[8] J. C. Bose was a close friend of Rabindranath and visited him often in Calcutta and in Santiniketan, where the Tagores started a university. Bose was educated at St. Xavier's College in Calcutta and went to England to study at Cambridge University, eventually returning to Calcutta to set up the Bose Research Institute in 1917. Bose's research into microwave physics was readily accepted and used by his European contemporaries. In fact, it was Bose's mercury coherer that was used by the Italian scientist Gugliemo Marconi to receive the radio signal in his first transatlantic radio communication experiment. Yet his research into plant irritability or response was met with some hostility in the context of the Victorian mechanistic materialist philosophy of science. Prominent electrophysiologists at the time were reluctant to accept Bose's conclusions that all plants possess a nervous system, a form of intelligence, and a capacity for remembering and learning (Shepherd 2012). Bose's ideas did,

জগদীশের ধ্যানভঙ্গ।

Figure 2 Gaganedranath Tagore, "Inanimate Scream." From Tagore 1921.

however, attract neovitalists such as the biologist and urban planner Patrick Geddes, who lived and worked in India and who saw biology as crucial to metaphysical thought.

Gaganendranath Tagore, in whose paintings and caricatures J. C. Bose and his experiments prominently featured, was born into the illustrious Tagore family in 1867. His artist brother Abanindranath Tagore (1871–1951) and his polymath uncle Rabindranath Tagore were amply involved in the intellectual and creative fervor that characterized the early years of the Indian Freedom movement in elite Bengali circles. At the turn of the century, the Tagore household in Jorasanko, Calcutta, became a center for what the art historian Partha Mitter calls "cultural nationalism," with the constant presence of the British art teacher and reformer Ernest Binfield Havell (1861–1934) and Margaret Noble (Sister Nivedita), who was a disciple of Swami Vivekananda (1863–1902), among others (Mitter 1994: 219–66). This complex cultural milieu, which gave rise to modernism in Indian art, was also heavily involved with cognate social movements in education and religious reform and included key players in the gradual overhaul of colonial power. Gaganendranath saw Bose's experiments in plant sentience for what they were, part of the cultural, political, and religious life of late colonial India, not separate from it. Gaganendranath Tagore created "Inanimate Scream," a satirical drawing, featuring Jagadish Chandra Bose, which is contained in a portfolio of "satirical pictures" published in 1921 titled *Reform Screams: A Pictorial Review at the Close of the Year 1921* (Tagore 1921). The portfolio was a broader political commentary that drew on recent and unjust constitutional reforms implemented by the British government in India. Gaganendranath's drawing depicts Bose seated in the mountains. From his hand, a spark like an inverted thunder bolt resonates tiny waves across the landscape. The figure with its outline of curly locks and sparse facial features sports an obvious third eye on his forehead. Around him, the trees and plants are alive. Here you can see plants marching "on strike," asking for "chanda," a monetary subscription for a cause. In the foreground, smaller plants writhe and move. The flat leaves of the lotus here proclaim "vande mataram," the title of a poem in praise of the motherland composed by Bankim Chandra Chattopadhyay in the 1870s, which went on to become a politically active slogan in the struggle for Indian independence. Beside the lotus, the desmodium, also known as the telegraph plant, seems to move its leaves in a synchronized dance to a call to "agitate." On the far right, the mimosa plant (also known in Bengali as *lajjabati lata*, or the "shy plant") twists away from itself to a chant of "shame shame." Gaganendranath's image of Bose seated in the mountains with a third seeing eye conflates him with the Hindu deity Shiva, who resides in the Kailash mountain range. Associated with lightning and thunder, Shiva's third eye and trident stand in for the forces of creation and destruction. Importantly it makes the connection with the ways in which Bose conflated modern science and ancient Hindu thought. Gaganendranath's attention to detail also signals his interest in Bose's research. While the lotus (*Nelumbo nucifera*) activates a reference to Indian myth and culture, the desmodium and mimosa plants come straight out of Bose's research.

Bose's scientific research, and especially his experiments in plant physiology, was not divorced from ideas of life and living mechanisms in Indian philosophy but rather in direct conversation with these ideas. In 1896 and 1897, Bose was focused mainly on his work in physics; in

1900 and 1901, he began to grow more and more interested in plants. Between these periods there was a fundamental shift in the way in which Bose related his work increasingly toward "Eastern" spirituality and antimaterialist sentiment. The epigraph to Bose's (1902) first scientific monograph, *Response in the Living and Non-living*, reads, "The real is one: wise men call it variously." Quoting this well-known declaration from the Rig Veda made Bose's position clear. It publicly corroborated his beliefs that, as proposed by the Vedas, the animate and the inanimate world was one, and that his electrographic discovery served to scientifically prove it (Brown 2016: 104). At the International Congress of Physicists in Paris in 1900, the Hindu spiritual leader Swami Vivekananda heard Bose present a paper. Vivekananda, who preached the superiority of Vedic spiritual thought and revitalized the nondualism of the Upanishads (a set of texts found within the Vedas), saw Bose as the embodiment of a new India in which the world of contemporary physical research was at one with ancient spiritual thought. Bose's scientific stance was, therefore, soon to become a political one. Legitimizing science not simply as a knowledge system created and ratified by the West but rather as a discipline perfectly compatible with and perhaps bound to Eastern philosophy, his work set into motion a new kind of nationalism embraced and disseminated by political figures such as Rabindranath and Vivekananda. Not only did the Hindu nationalists co-opt Bose, it seems that to a large extent he too co-opted them, bolstering a reform-oriented Hindu nationalism that, while largely tolerant, imagined India primarily as a Hindu nation and did not account for the rich complexity of religious groups in India of that time. A letter written to Rabindranath in 1901 acknowledges Bose's commitment to the freedom struggle and demonstrates the links he made between biology, the philosophies of science, and colonial politics. "I am alive with the life force of the mother Earth," he writes, "I have prospered with the help of the love of my countrymen. For ages the sacrificial fire of India's enlightenment has been kept burning, millions of Indians are protecting it with their lives, a small spark of which has reached this country [Britain] (through me)" (Sen 1994: 92).

The nondualistic, monistic philosophy that Vivekananda espoused emphasized the themes of oneness, unity, identity, and liberation. It sought to move philosophical insight and spiritual practices away from the established hierarchy in Hinduism of seers and yogis to more quotidian practices that gave spiritual power to the people (Boundas 2009: 649). In 1906, Ramananda Chatterjee (1865–1943) of the *Modern Review* described Bose's *Plant Response* as the greatest work of the Swadeshi movement—surpassing the burning of foreign clothes and the establishment of national universities (Chakrabarti 2004: 202). Across India in the late nineteenth and early twentieth centuries, Hindu intellectuals pushed the idea of monotheistic Hinduism, and this gained new reinforcement from a proposed indivisibility of science and religion. "The influence of positivism," according to historian Gyan Prakash (1999: 72), "was palpable in this idea, and positivist philosophers were often cited to legitimate dispositions that, according to Hindu intellectuals, Hinduism itself contained." Bose's discoveries provided an obvious and easy link to the sciences, until then seen by Europe as an exclusive domain of superiority, named and owned by Europeans (Chakrabarti 2004: 186–87). Bose's plant experiments became, in short, a nationalist strategy.

As admirable and productive as some of these nationalist strategies were for India's anticolonial struggle in the 1920s and 1930s, this nevertheless demonstrates the co-optability of plant science into religious—in this case, Hindu—ideology. Especially considering how Hindu religion and ideology have, in the twenty-first century, been critical forces in rewriting history and overriding constitutional rights in pursuit of a muscular and intolerant Hinduism (Subramaniam 2019), we ought to take note that this capacity for co-option runs serious risks in the wrong hands.

On the surface, the nondualism of Advaita Vedanta (that Vivekananda and his circle propagated) would seem to advocate for an "ecological consciousness," a sense of empathy with the natural world. This is based on the assumption that the nature of the self in Hinduism includes all lesser forms of existence (Crawford 1982; Deutsch 1970). However, scholars such as Lance Nelson and others argue that while Advaita fosters values such as simplicity of life, frugality, and nonviolence, in emphasizing transcendence it tends to devalue and neglect lower beings and the natural world. When understood in the broadest possible (and reductive) way, even in Hinduism, Buddhism, Jainism, and adjacent philosophies that originated in South and Southeast Asia, sentience is a metaphysical quality that all nonhumans possess, thus making all things deserving of respect and care. While these religions have been celebrated as being more "ecological" in their outlook, the knowledge of plant sentience, as this essay will show, does not automatically lead to an ethics of justice oriented toward plant life, or nonhuman life more broadly.[9] Therefore, although Bose's conclusions augmented Hinduism's general principles and validated the Hindu assumption that nonhumans, including plants, are sentient, it did not directly lead to conversations around nonhuman ethics or justice.

This particular moment in the early 1920s is an important one both for plant sentience studies and for anticolonial nationalism in India. The drawing and the nexus of characters I describe are testament to plant science's entanglement with larger colonial and nationalist histories. What this demonstrates, in particular, is that even when debated and discussed in the cultural and religious spheres, plant sentience or recognition of the agency of plants do not necessarily lead to a consideration of multispecies ethics or justice but can simply circle back to and be assimilated into contemporary human needs and hegemonic ideologies.[10]

Plants, Humans, and Colonial Histories

I started this essay by showing how projects in which humans apparently turn toward beings other than humans are always mediated by existing human-centered discourses and relationships with colonial and economic world orders. I conclude by reconsidering the complex relationship between plants, humans, and their entangled colonial histories. I argue that to move beyond a relationist ontology to a framework of ethics and justice, it is important to consider plant sentience alongside colonial and extractive histories. To do this, I return to Gaganendranath's satirical picture.

There is something else happening in Gaganendranath Tagore's satirical picture that warrants attention. The dancing plants on protest marches and singing the National Anthem suggest that there is a slippage between plant subject and colonial subject. This is in keeping with the fact that Gaganendranath's satirical works were especially invested in overt anticolonial critique. In the drawing, the plants dance as if in the power of an external force.

While their moves are supplemented with political slogans, their inability to act autonomously fulfills the pathos and self-irony that likens the plant subjects to colonized Indian subjects at the turn of the century. There is a revolution waiting to happen in the picture, as there was in India's political life at the time. Yet, as Tagore implies, the colonial political subjects, much like the dissident plants, were stalled at that moment in a state of semiautonomy. India would not gain independence for another twenty-five years after Gagagendranath's drawing was made, and the 1920s in India saw divergent views on anticolonial resistance. Some visionaries, such as Gaganendranath's uncle Rabindranath Tagore, pursued a softer cultural nationalism that did not necessarily advocate for an immediate overthrowing of British imperialism. While it is Bose's experiment that argues for plants as agential beings, the analogy that Gaganendranath Tagore makes between oppressed plant subjects and human subjects requires further discussion.

There is a long tradition of establishing a relation of analogy between human domination and the domination of nature by highlighting similarities between two practices and between modes of thinking and classification associated with them. This is what Marjorie Spiegel ([1988] 1996), in reference to slavery, has called "the Dreaded Comparison." Spiegel claims that like human slaves, nonhuman animals are subjected to branding, restraints, beatings, auctions, the separation of offspring from their parents, and forced voyages. Pointing to these similarities has sometimes been highly effective in illuminating some of the tropes of oppression wherein some humans are relegated to the natural and transactional sphere (Hage 2017; Yusoff 2018). However, a simple comparison erases the situated histories within which oppressions form.

Plants have been an extremely effective tool for colonial and economic domination from early on in colonial endeavors. As Alfred Crosby (1986) famously observed in *Ecological Imperialism*, plants and our relationship to them are literally rooted in culture and history. Botanical gardens in Europe are underwritten by the space of the plantation, in that they served laboratories for mass agricultural production the exploitation of resources (human, animal, vegetal, mineral) (Drayton 2000; Schiebinger 2007). Botany and the classification of forms of life enacted by imperial science need to be understood in the context of colonialism: "fathers" of botany such as Carl Linnaeus and Hans Sloane were significant figures in the development of scientific racism, the endorsement of slavery, and colonial expansion (Delbourgo 2007).

The contemporary artist Pedro Neves Marques's video *The Pudic Relation between Machine and Plant* (2016), while made in a very different time period and context from Gaganendranath's satirical picture, reflects on the conflation of human and plants as colonial subjects. In 2019, at the José de Guimarães International Arts Centre in Guimarães, Marques's video work opened the exhibition *Plant Revolution!* Featuring the plant *Mimosa pudica*, the video shows a looped interaction between a robotic hand and a mimosa plant. Recalling that *Mimosa pudica* derives its name from Carl Linnaeus's sexual taxonomy of plants, in which *pudica* refers to both the external sexual organs, shyness, and modesty, the artwork constructs an uncomfortable visual experience in which the repeated touching and shrinking of the mimosa plant references environmental colonialism and the inherent violence of colonial knowledge production. Linnaean taxonomy was central to the construction of race and sexuality. While in Marques's video, the plant is touched over and over again, the video is looped, and in reality, the plant was only touched a few times, not enough times for it to be conditioned against closing its leaves when the robotic arm came close. Using the tripartite juxtaposition of sexuality, technology, and science, the video urges the viewer to reconsider not only what plants feel but also longer histories of domination and asymmetrical power relations entangled in plant-human relations.

As this essay shows, plant sentience is a field rife with cultural and political manifestations. The two distinct strands of political co-option and biomimicry that I have chosen to highlight stand to serve as examples of the messy and complex ways in which ideas around multispecies lifeworlds are co-opted, repurposed, and directed to different ends. It brings to light a collision of often well-intentioned intellectual ambitions and calls to action around the question of human and nonhuman life, both of which are inherently rendered political. The examples demonstrate that a deep interest and engagement with plant sentience does not directly map on to any ethical or political outcomes. It does not in itself lead to legal subjecthood for plants or a sustained conversation about multispecies justice. The recognition of ontological subjectivity remains inherently co-optable. To remain acutely aware of these histories and current conditions of co-optability is the first step for artistic practice and cultural theory. It is this awareness that, I hope, will open up a broader range of possibilities and structures toward pluriversal approaches to multispecies ethics and justice.

Notes

1. Bose's 1913 monograph *Researches on Irritability of Plants* follows Francis Glisson and Albrecht von Haller's use of the term *irritability*. Alfred von Haller's (1801) experiments on animals led him to make the distinction between "sensibility"—the ability to perceive a stimulus—and "irritability"—the ability to respond to that stimulus. Bose's conclusion is as follows: "In surveying the response of living tissues we find that there is hardly any phenomenon of irritability observed in the animal which is not also found in the plant" (Bose 1913: 360).
2. In 1914, Bose lectured widely in Europe and America on plant irritability. At Cambridge, the botanist Francis Darwin, known for early research on the movement of plants with his father Charles Darwin, chaired Bose's lecture (Geddes 1920).
3. Examples include Creo Nova's interactive plant installation *The Genesis of Biosynthia* and Danish artist Sebastian Frisch's *Biophonic Garden*, among others. See also Berrigan 2014.
4. Most recently, a group of contemporary plant scientists, while agreeing wholly with Bose's conclusions, have chosen to use the terms *phytoneuron* to refer to sieve elements carrying an electric current and *phytoneurology* for the general subject area (Calvo, Sahi, and Trewavas 2017). Philosopher Emanuele Coccia (2018: 126) also observes a need to take a step back from the analogizing tendencies prevalent in some quarters of the plant sciences—or what he describes as the "stubborn attempt to 'rediscover' organs 'analogous' to those that make perception possible in animals without trying at all to imagine . . . another possible form of the existence of perception, another way of thinking the relation between sensation and body."
5. The Sydney-based evolutionary ecologist Monica Gagliano, who, as a young researcher, worked at Stefano Mancuso's International Laboratory for Plant Neurobiology in Florence, performed a somewhat similar experiment with very different ends more than five years before the exhibition. Gagliano chose fifty-six potted mimosa plants (known for its propensity to close its fern-like leaves at the human touch) and created a structure that would allow them to drop from a height of fifteen centimeters every five seconds into a bed of foam. The plants were systematically dropped sixty times during each session. The results indicated that while the mimosa plants closed their leaves when dropped, after about only four, five,

or six drops, they began to reopen their leaves, and by the end of the sixty drops, their leaves were completely open. On the basis of these findings, Gagliano argued that like animals, plants acquire, and adjust themselves in response to, information from their environment. The mimosa plant's behavior in response to repeated physical disturbance exhibits clear habituation, suggesting some elementary form of learning. Having allowed the plants to rest and then repeating the experiment a month later, Gagliano concluded that mimosa can display the learned response even when left undisturbed in a more favorable environment for a month. See Gagliano et al. 2014.

6. Heather Paxson's (2008) conception of microbiopolitics introduced the idea of "human encounters with the vital organismic agencies of bacteria, viruses, and fungi."
7. See, for example, texts advocating for a new, green economy such as Hawken, Lovins and Lovins 1999; Braungart and McDonough 2002; Krupp and Horn 2008.
8. The scientific community remains adamantly split over whether plant behavior can be compared to animal behavior, given the former's lack of a brain or nervous system. The critical point of contention in this debate is not the scientific data produced by Bose or more recent experiments but in fact the terminology used to describe plant actions.
9. It is important to add that unlike these more established religions and philosophies, where philosophy may not extend readily into practice, various Indigenous cultures continue to support coconstitutive relationships between the land and its diverse human and more-than-human dwellers (Rose 2011; TallBear 2015).
10. For a comparative study that focuses on the lethal agency of invasive plants in South Africa and their co-option into apartheid racial and anti-immigrant ideologies, see Comaroff and Comaroff 2001.

References

Agamben, Giorgio. 2004. *The Open: Man and Animal.* Stanford, CA: Stanford University Press.

Alpi, Amedeo, et al. 2007. "Plant Neurobiology: No Brain, No Gain?" *Trends in Plant Science* 12, no. 4: 135–36. https://doi.org/10.1016/j.tplants.2007.03.002.

Barua, Maan. 2019. "Animating Capital: Work, Commodities, Circulation." *Progress in Human Geography* 43, no. 4: 650–69. https://doi.org/10.1177/0309132518819057.

Benyus, J. M. 1997 *Biomimicry: Innovation Inspired by Nature.* New York: Morrow.

Berrigan, Caitlin. 2014. "Life Cycle of a Common Weed." In *The Multispecies Salon*, edited by Eben Kirksey, 164–84. Durham, NC: Duke University Press.

Blaser, Mario, Marisol de la Cadena, and Arturo Escobar. 2013. "Introduction: The Anthropocene and the One-World." In *Pluriverse Studies Reader.* Unpublished manuscript.

Bose, Jagadish Chandra. 1902. *Response in the Living and Non-Living.* London: Longmans, Green.

Bose, Jagadish Chandra. 1921. *Reform Screams, Satirical Pictures.* Calcutta: Thacker and Spink.

Bose, Jagadish Chandra. 1926. *The Nervous Mechanism of Plants.* London: Longmans, Green.

Bose, Jagadish Chandra. 1913. *Researches on Irritability of Plants.* London: Longmans, Green.

Boundas, Constantin V., ed. 2009. *Columbia Companion to Twentieth-Century Philosophies.* New York: Columbia University Press.

Braun, Bruce. 2009. "Nature." In *A Companion to Environmental Geography*, edited by Noel Castree, David Demeritt, Diana Liverman, and Bruce Rhoads, 19–36. London: Blackwell.

Braungart, Michael, and William McDonough. 2002. *Cradle to Cradle: Remaking the Way We Make Things.* New York: North Point.

Brenner, Eric D., Rainer Stahlberg, Stefano Mancuso, Jorge Vivanco, František Baluška, and Elizabeth Van Volkenburgh. 2006. "Plant Neurobiology: An Integrated View of Plant signaling." *Trends in Plant Science* 11, no. 8: 413–19. https://doi.org/10.1016/j.tplants.2006.06.009.

Brown, C. Mackenzie. 2016. "Jagadish Chandra Bose and Vedantic Science." In *Science and Religion: East and West*, edited by Yiftach Fehige, 104–22. New York: Routledge,.

Calvo, Paco, Vaidurya Sahi, and Anthony Trewavas. 2017. "Are Plants Sentient? *Plant, Cell, and Environment* 40, no. 11: 2858–69. https://doi.org/10.1111/pce.13065.

Chakrabarti, Pratik. 2004. *Western Science in Modern India: Metropolitan Methods, Colonial Practices.* Delhi: Permanent Black.

Chamovitz, Daniel. 2012. *What a Plant Knows*. London: Oneworld.

Coccia, Emanuele. 2018. *The Life of Plants: A Metaphysics of Mixture*. London: Polity.

Comaroff, Jean, and John L. Comaroff. 2001. "Naturing the Nation: Aliens, Apocalypse, and the Postcolonial State." *Social Identities* 7, no. 2: 233–65.

Crawford, S. Cromwell. 1982. *The Evolution of Hindu Ethical Ideals*. Honolulu: University of Hawai'i Press.

Crosby, Alfred. 1986. *Ecological Imperialism: The Biological Expansion of Europe, 900–1900*. Cambridge: Cambridge University Press.

Darwin, Charles. 1880. *The Power of Movement in Plants*. Assisted by Francis Darwin. London: John Murray.

Delbourgo, James. 2007. *Science and Empire in the Atlantic World*. New York: Routledge.

Deutsch, Eliot. 1970. "Vedānta and Ecology." *Indian Philosophical Annual* 7: 79–88.

Drayton, Richard Harry. 2000. *Nature's Government: Science, Imperial Britain, and the "Improvement" of the World*. New Haven, CT: Yale University Press.

Ellis, Erle C. 2013. "Sustaining Biodiversity and People in the World's Anthropogenic Biomes." *Current Opinion in Environmental Sustainability* 5: 368–72. https://doi.org/10.1016/j.cosust.2013.07.002.

Ernwein, Marion, Franklin Ginn, and James Palmer, eds. 2021. *The Work That Plants Do: Life, Labour, and the Future of Vegetal Economies*. Bielefeld, Germany: Transcript Verlag.

Forlano, Laura. 2016." Decentering the Human in the Design of Collaborative Cities." *Design Issues* 32, no. 3: 42–54. https://doi.org/10.1162/DESI_a_00398.

Gagliano, Monica. 2013. "Green Symphonies: A Call for Studies on Acoustic Communication in Plants." *Behavioral Ecology* 24, no. 4: 789–96. https://doi.org/10.1093/beheco/ars206.

Gagliano, Monica, Michael Renton, Martial Depczynski, and Stefano Mancuso. 2014. "Experience Teaches Plants to Learn Faster and Forget Slower in Environments Where It Matters." *Oecologia* 175, no. 1: 63–72. https://doi.org/10.1007/s00442-013-2873-7.

Galansino, Arturo. 2018. *The Florence Experiment: Un progetto di Carsten Höllen e Stefano Mancuso*. Venice: Marsilio.

Geddes, Patrick. 1920. *The Life and Work of Sir Jagadis C. Bose*. London: Longmans, Green.

Goldstein, Jesse, and Elizabeth Johnson. 2015. "View Biomimicry: New Natures, New Enclosures." *Theory, Culture, and Society* 32, no. 1: 61–81. https://doi.org/10.1177/0263276414551032.

Hage, Ghassan. 2017. *Is Racism an Environmental Threat?* Cambridge: Polity.

Hall, Mathew. 2011. *Plants as Persons: A Philosophical Botany*. Albany: SUNY Press.

Haller, Alfred von. 1801. *First Lines of Physiology*. Edinburgh: Bell and Bradfute.

Hamilton, Jennifer. 2016. "Bad Flowers: The Implications of a Phytocentric Deconstruction of the Western Philosophical Tradition for the Environmental Humanities." *Environmental Humanities* 7, no. 1: 191–202. https://doi.org/10.1215/22011919-3616398.

Haraway, Donna. 2003. *The Companion Species Manifesto: Dogs, People, and Significant Otherness*. Chicago: Prickly Paradigm.

Haraway, Donna. 2008. *When Species Meet*. Minneapolis: University of Minnesota Press.

Hauter, Wenonah. 2014. *Foodopoly: The Battle over the Future of Food*. New York: New Press.

Hawken, Paul, Amory B. Lovins, and L. Hunter Lovins. 1999. *Natural Capitalism: Creating the Next Industrial Revolution*. London: Earthscan.

Hayles, N. Katherine. 2002. "Flesh and Metal: Reconfiguring the Mindbody in Virtual Environments." *Configurations* 10, no. 2: 297–320.

Johnson, Elizabeth R. 2010. "Reinventing Biological Life, Reinventing 'the Human.'" *Ephemera* 10, no. 2: 177–93.

Johnson, Elizabeth, et al. 2014. "After the Anthropocene: Politics and Geographic Inquiry for a New Epoch." *Progress in Human Geography* 38, no. 3: 439–56. https://doi.org/10.1177/0309132513517065.

Kahn, Douglas. 2013. *Earth Sound Earth Signal: Energies and Earth Magnitude in the Arts*. Berkeley: University of California Press.

Kareiva, Peter, Michelle Marvier, and Robert Lalasz. 2012. "Conservation in the Anthropocene: Beyond Solitude and Fragility." *Breakthrough Journal*, no. 2. https://thebreakthrough.org/journal/issue-2/conservation-in-the-anthropocene.

Kaufmann, Frederick. 2012. *Bet the Farm: How Food Stopped Being Food.* Hoboken, NJ: Wiley.

Kimmerer, Robin Wall. 2013. *Braiding Sweetgrass: Indigenous Wisdom, Scientific Knowledge, and the Teachings of Plants.* Minneapolis: Milkweed Editions.

Kohn, Eduardo. 2013. *How Forests Think: Toward an Anthropology beyond the Human.* Berkeley: University of California Press.

Krupp, Fred, and Miriam Horn. 2008. *Earth: The Sequel: The Race to Reinvent Energy and Stop Global Warming.* New York: W. W. Norton.

Marder, Michael. 2012. *Plant-Thinking: A Philosophy of Vegetal Life.* New York: Columbia University Press.

Mazzolai, Barbara. 2017. "Plant-Inspired Growing Robots." In *Soft Robotics: Trends, Applications, and Challenges*, edited by Cecilia Laschi, Jonathan Rossiter, Fumiya Iida, Matteo Cianchetti, and Laura Margheri, 57–73. Cham, Switzerland: Springer International.

Mazzolai, Barbara. 2019. "Soft Robotics." *Biomimetics* 4, no. 1. https://doi.org/10.3390/biomimetics 4010022.

Mazzolai, Barbara, and Cecilia Laschi. 2020. "A Vision for Future Bioinspired and Biohybrid Robots." *Science Robotics* 5, no. 38. https://doi.org/10 .1126/scirobotics.aba6893.

Mazzolai, Barbara, and Stefano Mancuso. 2013. "Smart Solutions from the Plant Kingdom." *Bioinspiration and Biomimetics* 8, no. 2. https:// doi.org/10.1088/1748-3182/8/2/020301.

Mazzolai, Barbara, Francesca Tramacere, Isabella Fiorello, and Laura Margheri. 2020. "The Bio-Engineering Approach for Plant Investigations and Growing Robots. A Mini-Review." *Frontiers in Robotics and AI* 7: 573014. https://doi.org /10.3389/frobt.2020.573014.

Miller, Elaine. 2002. *Vegetative Soul: From Philosophy of Nature to Subjectivity in the Feminine.* Albany: SUNY Press.

Mitter, Partha. 1994. *Art and Nationalism in Colonial India, 1850–1922: Occidental Orientations* Cambridge: Cambridge University Press.

Must, Indrek, Eduardo Sinibaldi, and Barbara Mazzolai. 2019. "A Variable-Stiffness Tendril-Like Soft Robot Based on Reversible Osmotic Actuation." *Nature Communications*, no. 10: 344. https://doi .org/10.1038/s41467-018-08173-y.

Myers, Natasha. 2017. "From the Anthropocene to the Planthroposcene: Designing Gardens for Plant/ People Involution." *History and Anthropology* 28, no. 3: 297–301.

Nealon, Jeffrey T. 2016. *Plant Theory: Biopower and Vegetable Life.* Stanford, CA: Stanford University Press.

Nordhaus, Ted, and Michael Shellenberg, eds. 2011. *Love Your Monsters: Postenvironmentalism and the Anthropocene.* Oakland, CA: Breakthrough Institute.

Ogden, Laura A., Billy Hall, and Kimiko Tanita. 2013. "Animals, Plants, People, and Things: A Review of Multispecies Ethnography." *Environment and Society: Advances in Research* 4: 5–24. https:// doi.org/10.3167/ares.2013.040102.

Oliver, Kelly. 2007. "Stopping the Anthropological Machine: Agamben with Heidegger and Merleau-Ponty." *PhaenEx* 2, no. 2: 1–23.

Paxson, Heather. 2008. "Post-Pasteurian Cultures: The Microbiopolitics of Raw Milk Cheese in the United States." *Cultural Anthropology* 23, no. 1: 15–47.

Perkins, Harold. 2007. "Ecologies of Actor-Networks and (Non)social Labor within the Urban Political Economies of Nature." *Geoforum* 38, no. 6: 1152–62.

Pettman, Dominic. 2013. "The Noble Cabbage: Michael Marder's *Plant-Thinking*." *Los Angeles Review of Books*, July 28. https://lareviewofbooks.org /review/the-noble-cabbage-michael-marders -plant-thinking.

Plumwood, Val. 2002. *Environmental Culture: The Ecological Crisis of Reason.* London: Routledge.

Pollan, Michael. 2013. "The Intelligent Plant: Scientists Debate a New Way of Understanding Flora." *New Yorker*, December 23–30, 92–105.

Prakash, Gyan. 1999. "Science Between the Lines." In *Subaltern Studies X: Writings on South Asian History and Society*, edited by Gautam Bhadra, Gyan Prakash, and Susie Tharu, 59–82. Oxford: Oxford University Press.

Rajan, Kaushik Sunder. 2006. *Biocapital: The Constitution of Postgenomic Life.* Durham, NC: Duke University Press.

Robin, Marie-Monique, dir. 2008. *The World According to Monsanto.*

Rose, Deborah Bird. 2011. *Wild Dog Dreaming: Love and Extinction.* Charlottesville: University of Virginia Press.

Rysman, Laura. 2018. "An Artist's Modern Slides: Inside a Fifteenth-Century Palazzo." *New York Times*, April 20.

Sadeghi, Ali, Alessio Mondini, and Barbara Mazzolai. 2017. "Toward Self-Growing Soft Robots Inspired by Plant Roots and Based on Additive Manufacturing Technologies." *Soft Robotics* 4, no. 3: 211–23. https://doi.org/10.1089/soro.2016.0080.

Schiebinger, Londa. 2007. *Plants and Empire: Colonial Bioprospecting in the Atlantic World.* Cambridge, MA: Harvard University Press.

Sen, Dibakar. 1994. *Patrabali Acharya Jagadish Chandra Bose.* Calcutta: Bose Institute.

Shepherd, Virginia Anne. 2009. *Remembering Sir J. C. Bose.* Edited by D. P. Sen Gupta and Meher H. Engineer. Hackensack, NJ: World Scientific.

Shepherd, Virginia Anne. 2012. "At the Roots of Plant Neurobiology." In *Plant Electrophysiology*, edited by Alexander G. Volkov, 3–43. Berlin: Springer.

Spiegel, Marjorie. (1988) 1996. *The Dreaded Comparison: Human and Animal Slavery.* New York: Mirror.

Subramaniam, Banu. 2019. *Holy Science: The Biopolitics of Hindu Nationalism.* Seattle: University of Washington Press.

Sundberg, Juanita. 2014. "Decolonizing Posthumanist Geographies." *Cultural Geographies* 21, no. 1: 33–47. https://doi.org/10.1177/1474474013486067.

Sundberg, Juanita. 2015. "Ethics, Entanglement, and Political Ecology." In *The Routledge Handbook of Political Ecology*, edited by Thomas Perreault, Gavin Bridge, and James McCarthy, 117–26. New York: Routledge.

Tagore, Gaganendranath. 1921. *Reform Screams: A Pictorial Review at the Close of the Year 1921.* Calcutta: Thacker, Spink.

TallBear, Kim. 2015. "Theorizing Queer Inhumanisms: An Indigenous Reflection on Working Beyond the Human/Not Human." *GLQ* 21, nos. 2–3: 230–35.

Todd, Zoe. 2015. "Indigenizing the Anthropocene." In *Art in the Anthropocene: Encounters Among Aesthetics, Politics, Environment and Epistemology*, edited by Heather Davis and Etienne Turpin, 241–54. London: Open Humanities Press.

Trewavas, Anthony A. 2006. "A Brief History of Systems Biology: 'Every Object That Biology Studies is a System of Systems.' Francois Jacob (1974)." *Plant Cell* 18, no. 10: 2420–30. https://doi.org/10.1105/tpc.106.042267.

Viveiros de Castro, Eduardo. 2012. *Cosmological Perspectivism in Amazonia and Elsewhere.* Hau Masterclass Series 1. Manchester: HAU, 2012.

Wandersee, James H., and E. E. Schussler. 2001. "Toward a Theory of Plant Blindness." *Plant Science Bulletin* 47, no. 1: 2–9.

Wohlleben, Peter. 2016. *The Hidden Life of Trees: What They Feel, How They Communicate—Discoveries from a Secret World.* Vancouver, BC: Greystone.

Wolfe, Cary. 2010. *What Is Posthumanism?* Minneapolis: University of Minnesota Press.

Yusoff, Kathryn. 2018. *A Billion Black Anthropocenes or None.* Minneapolis: University of Minnesota Press.

Sria Chatterjee is an art historian and environmental humanities scholar. She is the head of research and learning at the Paul Mellon Centre for British Art in London. She specializes in the political ecologies of art and design from the colonial to the contemporary and leads the multiyear research project Climate and Colonialism at the Paul Mellon Centre. Sria received her PhD from the Department of Art and Archaeology at Princeton University in 2019 and has held fellowships and research positions at the Max-Planck Kunsthistorisches Institute and the Swiss National Science Foundation. She has published numerous chapters, essays, and articles in *British Art Studies, Museums History Journal, Contemporary Political Theory*, and other outlets, including *Noema Magazine*. Her first book, currently in progress, provides a close look at the deep links between nationalism, agriculture, and the natural environment through the history of art, design, and media. In 2021, Sria started *Visualizing the Virus*, which investigates the diverse ways in which pandemics are visualized and the inequalities they make visible.

OCEAN JUSTICE

Reckoning with Material Vulnerability

Susan Reid

Abstract **The continued campaign of violence by extractivists against multibeing relations, embodied beings, and ecological living is bewildering. Coded by mastery, and as a carrier of its values, international laws of the sea facilitate these campaigns by legitimating ecological abuse. As such, responding to the ocean's declining conditions with more laws and regulations alone misses how underlying cultural values contribute to the production of ecological harm. This article considers how the imaginary of mastery underpinning dominant ocean governance regimes enables the production and distribution of vulnerability. Thinking with the ocean reveals how anthropogenic harms manifest and proliferate both materially and through the discursive networks of ocean governance. Though material vulnerability is a condition that brings us into being interconnectedly with other worlds, it also (unevenly) implicates us in ocean harm. This article draws on feminist posthumanist, legal, and marine scientific work to examine these issues in the context of an emerging concept of ocean justice, in which the conditions for cohabiting well with the seas might be imagined and activated.**

Keywords **justice, ocean, multibeing, multispecies, vulnerability, extractivism**

At the rim of a small rock pool, fringes of sepia algae swoon in the wash of an ebbing tide. Tufts of Neptune's necklace cling to the inner shelves, jostling for space between encampments of hunkered limpets, whelks, and black nerites. Having out-sustained most of the colorful, fleshy anemones, the stubborn sessile ones have inherited the rockpool. As a regular visitor to these coastlines, I've seen the jewel-like worlds in these lithic portals become monochrome bunkers of toughened

Cultural Politics, Volume 19, Issue 1, © 2023 Duke University Press
DOI: 10.1215/17432197-10232516

holdfasts and gelatinous algae. It is hard not to imagine these browning pools as missives from a troubled ocean. Over just a few decades, too many beautiful coastal ecotones seem to be very quietly emptying. Those who live with the sea notice more plastic crumbs, fewer pipis burrowing the wet sands, fewer shell treasures tangled in wrack, and miss the terns and sandpipers scampering from the waves.

1. Imagining the Possibility of Cohabitating Well with *the Seas*

Defining ocean justice is like trying to catch the Gulf Stream with a cast net. Familiar logics seem wholly unsuitable to the task. Nevertheless, some provisional notions of ocean justice are called for because the extractive activities and related sonic, heat, plastic, and chemical waste streams of powerful corporations are transforming the very constitution of the seas.[1] These destructive interests are validated by an ocean governance regime that supports a resource extraction model of development and a cultural imaginary of mastery unable to register this as a problem. The overarching legal framework is the United Nations Convention on Law of the Sea,[2] an international agreement still caught in the undertow of Western imaginaries of mastery and therefore inadequate to the task of enabling ways of living well with the ocean. At the same time, responsibility for the ocean's declining conditions cannot be shifted entirely to this regime or the influence of imaginaries of mastery and extractivism—individual and collective material entanglements also contribute. A "politics of purity" (Shotwell 2016) is just not possible, nor indeed is it desirable. We are individually co-constituted with the world and, as Alexis Shotwell (2016: 10) writes, "contaminated and affected." The conditions of material embodiment and vulnerability, so immanent to being, connect us with human and more-than-human others. They also implicate us (albeit unevenly) in the anthropogenic activities that are changing ocean chemistries and flows (IPCC 2019). We are all transitioning with these material flows, but how they will impact still-to-come human and more-than-human lives and lifeways is unclear—not least because much of the marine realm is physically and ontologically unknowable.[3] This is the context in which my concept of ocean justice unfolds as a concern to imagine possibilities for cohabiting well with the seas.[4] In this essay I explore the conceptual spaces that ocean justice could occupy as well as some of its epistemic, ontological, and political entanglements and potentials.

How might responsible cohabitation with the ocean be approached in ways that enable them[5] to just "be"? This is the modest aspiration suggested by Stacy Alaimo (2019: 409n6) as an acknowledgment that anthropogenic changes already impacting the planet may have foreclosed the possibility for biodiverse flourishing. I agree with this humble goal. Hoping for a situation where the ocean could *just be* reflects how habitability is increasingly contingent for marine beings experiencing the ocean's rapidly changing material, sonic and biodiverse conditions. For so long as humans intervene in ocean realms, the scale and nature of our extractions and waste excesses will influence the ocean's capacity for continual becoming or bare survivance. As well, given the indeterminate spatial and temporal scale of the ocean's transitions, any formulation of ocean justice can only be provisional and at times speculative, and needs to be always open to revision. Importantly, the question of responsible cohabitation with the ocean is not exclusively a matter of needing better law (though stronger and wider conservation provisions would help), or even more marine science (as much as scientific knowledges are invaluable). Rather, it is a cultural matter that concerns different perceptions and relationships with the ocean and the concepts of subjectivity and humanity that inform these relations.

Since UNCLOS came into force in 1994, the extent of extractivist violences across and beneath the seas has increased. As the convention controls humanity's dominant relationship with the seas, it is reasonable to assume that it is failing both the ocean and us. This failure warrants new tools to reimagine our relations otherwise. My proposition for an ocean justice framework is a response to the scale of extractivism's unrelenting drawdown on the ocean and the complicity and conservation failures of the ocean governance regime.[6] It is a work in progress that sits outside Western legal frameworks, taking form athwart and beneath law, bumping at its edges, nudging its set (normative) course, giving notice of omissions and occasionally boring holes at the sides to allow the ocean to wash through and into legibility. The point is to disrupt the ideological potency with which international ocean law and its institutions interfere in the possibility of cohabitating well with the seas.

Imaginative knowledge-building approaches and ways of thinking with the seas are needed to both attune to the ocean and discern how governance frameworks produce and distribute harms, including by their exclusions of the ocean. Ocean justice engages in "seatruthing"—a term I have coined to describe practices that test the seaworthiness of claims made by legal frameworks, institutions, and industry and other discourses or commodities where the ocean is implicated but not unacknowledged. Seatruthing makes no claims for singular, rarefied notions of truth but, rather, notices and interrogates what these representations do when brought into relation with different oceanic elements.[7] Seatruthing interrogates where and how the ocean is missing in such accounts and seeks to understand how these absences create ecological harm. The practice recruits multiple and diverse testimonies of the ocean and their constituencies to activate the ocean's voice as an interlocuter on matters that may affect them. Drawing on multiple ocean voices can also reveal how anthropogenic harms traverse permeable material and discursive borders in ways that reverberate through biological and social communities and earth systems.

Ocean justice practices ought to be self-reflexive and open to extending concepts of ecological subjectivity to acknowledge the entanglements of embodiment and material vulnerabilities, and the inherent violences within human relations with the seas. Katherine Yusoff (2012) analyses the inherent violences within human and more-than-human relations and proposes that these violences be brought into visibility in order to eschew them. Though indebted to Yusoff's rich analysis, I argue that because of human material dependencies not all these violences can be eschewed. Humans require material provisions to feed our flesh bodies and the "prosthetics" through which we participate in the world: basic housing, information and communication technologies, and domestic commodities.[8] Our dependencies on these materials renders us vulnerable, and provisioning them entails violences that kill or harm more-than-human others and lifeways. I figure this provisioning for our biological bodies and prosthetics as material predation.

At the extremities of material predation, voracious extractivism attenuates conditions of livability for us all. Recalibrating subjectivity to recognize humans as both vulnerable and as material predators sits uncomfortably alongside more desirable concepts of multibeing kinships and ocean cohabitation. However, these discomforts can productively inform efforts to mitigate the degree and nature of violences against the ocean as well as ethical obligations and responses toward the worlds of our prey. Any useful notion of ocean justice needs to take these intersecting determinants of material predation and vulnerability into account.

Ocean justice differs from rights of nature laws and earth jurisprudence models in which entities such as a river or lake are granted legal personhood.[9] Valuable symbolically, rights of nature models encouragingly provide legal recognition of the cultural and spiritual significance of nonhuman natural subjects. However, the generalized descriptions employed by such models risk homogenizing other natures, eliding the complexities of material, living worlds and localized, particular vulnerabilities. As well, the concept of human envisaged in these models is, for the most part, underexamined.

As with rights of nature responses, a concept of ecological law is emerging within Western legal frameworks that is more sensitive toward nonhuman worlds.[10] Ecological law engages concepts of human interconnectedness and kinship with other nature (Sbert 2020: 77–78) and environmental governance considerations that respect an ecocentric, "mutually-enhancing human-Earth relationship" (78). Variations of these concepts are, of course, well developed within different cultural traditions, notably long-standing Indigenous practices of stewardship and kinship with other natures.[11] The concept of ecological integrity, which is central to ecological law, is difficult to align with the ocean's complex, changing conditions.[12] It is difficult to conceive an original ecological state for the ocean, given their continuous transformations. Today's ocean is not the same as that which existed when UNCLOS first commenced, and they are already tomorrow's past. Even if restoring ecological integrity were feasible for the ocean, the vast limitations of the Western scientific understanding of deep ocean realms unsettles any claims of reliably knowing what this concept could mean for the deep. As well, scholarly developments in the field of environmental law signal a shift toward posthumanist perspectives, such as the model of critical environmental law (CEL) advanced by Andreas Philippopoulos-Mihalopoulos (2011, 2017). Unlike CEL, ecological law, and the earth jurisprudence discussed here, my concept of ocean justice explicitly engages the ocean and turns to them as a source of knowledge-building.

2. A Worldview That Denies Vulnerability Exacerbates Vulnerability

The concept of vulnerability built into my ocean justice framework draws on ecological justice scholarship and feminist legal theory. In particular, Martha Fineman's "vulnerable legal subject" thesis emphasizes a concept of embodied and relational subjectivity that challenges law's paradigmatic, disembodied, *invulnerable*, "legal subject" (Fineman 2008). Fineman's thesis argues that vulnerability is experienced unevenly across life stages and temporal scales; and through particular circumstances, such as physical ability, livable environments and climate, and access to food, water, and social support. Though it conceptualizes vulnerability as an enduring aspect of the human condition, Fineman's thesis translates easily to nonhuman domestic animals, given their dependencies on humans (Satz 2009: 79). Applying a materialist reading to the concept's implicit embodiment as, others have done (see, for example, Grear 2015; Kotzé 2019: 84), enables the ocean, interdependent beings, and earth systems to be recognized as subjects exposed to the condition of vulnerability.

The concept of a vulnerable legal subject challenges the idealized subject of law. Feminist legal scholars argue that this idealized subject encounters life in a way that most closely aligns with a corporation (Naffine 2003; Grear 2015; see also Fineman and Grear 2013). Unlike embodied, vulnerable, and fleshy human subjects, corporations are most capable of fulfilling mastery's criteria of disembodiment, autonomy, and detachment. Corporations are the paradigmatic human of legal personhood (Grear 2015), and, faced with a challenging environmental situation or the depletion of resources, for example, they can liquidate, restructure, and/or relocate to more favorable conditions (Naffine 2003; Grear 2015). Hierarchical values of mastery also find their way into the processes of law. Anna Grear writes insightfully that the law is "a viral carrier of the hierarchy of being" (Grear 2015: 86–87), which goes some way to explain the inadequacy of international ocean laws' responses to marine ecological stresses. It has sedimented hierarchies of being (86–87) that prioritize humans over other natures and certain groups of humans over others (82–83). We can see this in the privileging of corporations and wealthy nations over communities and less wealthy nations.[13] Being alert to mastery's privileging, we can examine how it manifests in the implementation of legal devices such as the doctrinal legal principal of the common heritage of mankind.[14] Mining proponents use the principle to justify their extractive interests, with the argument that mining the seabed is in the interest of "mankind," or humanity. Taken at face value, we might reasonably assume that all humans have an interest in the seabed, however, in its legal tilt and implementation, the principle is biased toward corporate humanity.

The sway of mastery renders the ocean governance regime incapable of recognizing or responding adequately to the harm associated with extractive activities. The regime claims authority and advances oceanic representations based on a scintilla of empirical knowledge of the deep and remote ocean and most of their beings. With limited ocean knowledge, the original drafters of UNCLOS filleted the ocean into territorial zones of exploitation (such as the Area and Exclusive Economic Zone) that bear no resemblance to the actual ocean. Philip Steinberg and Kimberley Peters (2015: 249–50) write that such spatial partitioning "serves only to reveal [the ocean's] unknowability as an idealized, stable and singular object." It is this mythical version of the ocean that the convention purports to govern. In other words, not only does the convention claim its authority on a limited knowledge base, it also presides over an ocean simulacrum—a nonentity. By privileging extractive development, UNCLOS licenses harm. By operating with a paucity of knowledge and not *getting* the actual ocean, the convention produces and compounds vulnerability.

Imaginaries of mastery and their tributary "predatory ontologies" of Western extractive liberalism (Ruder and Sanniti 2019),[15] entrench and normalize exploitative relationships with human and more-than-human others. Under the convention's Enlightenment values, marine nature is put to work in the service of human enterprise. Together with other laws of the seas, the convention's legal representations and discursivities facilitate marine exploitations by spatializing and territorializing the ocean as a background place where certain human actions are legitimated: marine creatures harvested, commercial species protected, cables laid, bombs tested, shipping traffic crisscrossed, waste dumped, bedrock drilled, fossil fuels extracted. In short "the ocean is instrumentalised as quarry, pantry, sink and sump" (Reid 2020). While this compels ocean justice responses, this is not to overcome mastery's ideological stronghold. As Juliette Singh (2017) observes, mastery is just too pervasive for that to be practically possible. Rather, the point is to build and proliferate conceptual meanders and eddies around mastery through which living well with the ocean can be imagined and actioned.

Mastery produces vulnerabilities by failing to recognize material dependency as a vulnerability. It promotes a false immateriality by disavowing embodiment and obscuring the places, (human and other) beings and labors that constitute and produce our prosthetics and commodities. It fails to recognize that the materials we consume have a provenance or that what we combust potentially reenters certain worlds as toxic waste. These epistemic limitations make it difficult to fathom the depth and proliferation of vulnerabilities and exacerbate the potential for harm in several ways. First, mastery fails to recognize material embodiment, vulnerability, and interdependencies across human and more-than-human worlds. Second, it denies the inalienability of violence inherent within our relations with more-than-human others and the profound ethical responsibilities that this obliges—that is, the violence associated with material provisioning to feed our fleshy selves, familial relations (such as companion animals), and prosthetics. Third, mastery's denialism reifies and instrumentalizes nonhuman nature. Where ever human interventions into nonhuman worlds are measured against those abstract representations (as is the case with UNCLOS), the harm that they inflict will not be sufficiently recognized. Finally, the imbrications of each of these denials amplify the already existing, universal condition of vulnerability that Fineman and others recognize. Mastery backgrounds the vulnerabilities of nonhuman natures but also applies the same treatment to humans. A worldview that denies vulnerability exacerbates vulnerability.

Recalling Shotwell's caution against a politics of purity, I note that while imaginaries provide cultural atmospherics for how humans connect with more-than-human worlds, this does not diminish individual agency. Danielle Celermajer (2018: 134), when discussing the conditions that enable torture, argues that even if the surrounding cultural values support particular actions, humans are not automatons "whose behaviour is determined [entirely] by external forces." The analytical point Celermajer offers is to recognize that while imaginaries influence worldly relations, it still takes individual agency to choose to perpetuate violences against humans and others. Individual executives, officials, and politicians ultimately sign off on decisions that can violently impact marine communities and proliferate vulnerabilities (see Reuters 2012). Further, individual agency is a factor in consumer choices that influence demand for the flesh of marine beings and other materials. By the same logics, imaginaries, actions, and cultural values that envisage the possibility for cohabiting well with the ocean can be created and sustained through the generative agency of individuals thinking and imagining in multivocal knowledge habitats.

Tweaking the current governance frameworks but not revisiting their underlying and troubling imaginaries will not likely bring about the change needed for ocean justice. To invoke Audre Lorde (1984), different tools are needed to dismantle the ecologically risky, foundational imaginary of law. If the ocean's declining condition can be partly attributed to legal mechanisms designed to privilege extractive capitalist institutions, then as long as imaginaries of mastery continue to influence the law and those biases, relying exclusively on precautionary legal reform will not be sufficient. Take as an example, the International Seabed Authority's Draft Regulations on Exploitation of Mineral Resources in the Area.[16] The Draft Exploitation Regulations are intended to regulate mining in seabed areas beyond national jurisdiction and as environmental protections. From another view, they constitute one more regulatory tool to legitimate and legislate harm against the ocean.

The authority is mandated by UNCLOS to oversee the protection of the marine environment from the harmful effects of seabed exploitation.[17] To this end, it embeds the precautionary principle in its Draft Exploitation Regulations[18] to ensure that mining contractors take "cost-effective" measures to prevent "serious or irreversible damage" to the environment, even if there is little scientific evidence supporting that risk.[19] Contractors are meant to take the principle into account when assessing and managing potential risks to the marine environment. However, operational and physical factors limit how effectively the principle is to be implemented, including: the self-regulatory environment of the deep seabed mining regime; regulatory capture of the Seabed Authority by corporations;[20] and the physical difficulty of scrutinizing mining operations several kilometers below the sea surface. The Seabed Authority conveys the appearance of taking its environmental responsibilities seriously by including the precautionary principle in the draft regulations but fails to ensure that it is sufficiently operationalized. Corporations benefit by regulating their own activities, relatively out of sight from expert and public scrutiny. Deep-sea ecologies such as those in the Clarion Clipperton Zone, for which very little is known, will potentially be subject to "serious or irreversible damage."[21] The principle's inclusion in the regulations functions as an outward-facing patina of concern for ecological vulnerability that lends legitimacy to the actions of mining corporations.

At best, "affirmative strategies" such as legal principles are merely correctives, the sort of technical or administrative fixes that Nancy Fraser (2003) describes as merely adding to the volume of regulatory mechanisms. The technical legal fixes instituted by the authority's draft regulations facilitate development and do little to prevent harm to increasingly vulnerable marine worlds. Meaningful change calls for the sort of "transformative strategies" advocated by Fraser that can dismantle the underlying framework or root cause of particular issues (74–77). My concept of ocean justice calls for such transformative strategies that can change the nature of our dominant relationship with the ocean and extend the concept of ecological subjectivity that is brought into that relationship.

Reckoning with who and how we think we are in connection with the ocean is one such strategy. How do we attune to the ocean so that we might better respond to what cohabiting well with the ocean might entail? This does not mean denying the inherent violence entailed in material provisioning from the ocean. Rather, it calls for an imaginary that can recognize and respond ethically to that violence, in the context of both human and ocean vulnerabilities, and our desires to keep worlding together. Cohabiting well with the ocean relies on our capacity to imagine the vulnerability of embodied being. Being receptive to vulnerability is, as Maria Puig de La Bellacasa (2012: 206) writes, an "ethical stance in the practice of thinking." After all, writes Astrida Neimanis (2014: 22), "we can only act and respond to a world that we can also imagine." Additionally, as Donna Haraway (2016: 35) writes, "It matters what matters we use to think other matters with; it matters . . . what thoughts think thoughts, what descriptions describe descriptions. . . . It matters what stories make worlds, what worlds make stories." We might say, following Neimanis and Haraway, that for ocean justice, it matters what imaginaries we imagine with.

3. Imagining Vulnerability and Its Flows

To an extent, knowledge is consubstantial with ocean justice. Unfortunately, though, where knowledge goes, the exploitations of extractive capitalism have tended to follow. The discovery of manganese nodules on the abyssal plain, for example, opened the ocean's seabed to extractive interests. Marine scientific inquiries are critical to understand conditions of ocean vulnerability. They enable us to "know something of those other worlds and species—or, at least, know what is and is not available to our ways of knowing," as Sria Chatterjee and Astrida Neimanis write (Celermajer et al. 2020). For the most part, however, the ocean is unknown and unknowable in the context of Western empirical scientific knowledge. That said, limits to knowledge, and in particular those of Western orthodox science, do not preclude myriad other ways of understanding and ethically connecting with the ocean. Multiple and diverse cultural worldviews and intellectual practices, which may not have intimate empirical knowledge of deep-living realms, continue to practice relations of respect, care, reciprocity, and kinship with the seas. Ongoing processes of colonization have tended to exclude such contributions from ocean governance regimes (Jackson 1993). Nevertheless, we might rightly question the need for more of a particular type of knowledge for justice to guide our interactions with the ocean. For while much of the ocean may be empirically unknowable, the wager is that a good number of more-than-human others that may be insensible to us are not beyond the reach of our harms. We can still respond with care to the likelihood of vulnerability while also acknowledging unknowability with epistemic humility.

To think with the ocean as a subject of vulnerability requires imaginatively reaching beyond the terrestrial shelf edge of ontological and epistemological comforts. This rigorous imagination is critical to ocean justice. So too are multivocal approaches that draw on ocean voices, multiplicities of scientific and other epistemic sources, traditional knowledges, and insights from artists and philosophers, always alert to anthropocentric and terrestrial-centric limitations. Courage is also needed to acknowledge that predating on more-than-human marine others for material provisions inflicts harms against other worlds. Additionally, both imagination and humility are needed to develop ethical frameworks for predation practices in which care and respect can be extended to the worlds of our prey. This at least is one strategy for parsing the scale and degree of violences that might be accepted in the context of necessary human predations. In these times of extractivism's unrelenting ocean assaults, imagination must ebb as far back from the familiar shoreline as possible to feel the edges of new ecological subjectivities and ways of connecting materially and conceptually with the ocean. How might we "chang[e] the shape of the thinkable" (Lather 1993: 687) and resist inventing ocean relationships from within familiar humanist, terrestrial, and extractive frameworks?

Rather than attempting to bring ocean subjects or simulacra of the unknowable ocean into the familiar light of onto-epistemic comforts, we could try immersing with them into the dark, to imagine and give testimony to what we don't or can't know.[22] In the face of unknowability, imagination enables us to conceptualize strategies for losing our oxygen, ocular, and solar centric ontologies, so that we might respond more sensitively and ethically to ocean vulnerabilities; and contemplate what might be necessary conditions of cohabitation with the ocean, locally and at a planetary scale (Reid 2020, 2022). Absence of sunlight, or the dense pressure caused by cubic kilometers of moving seawater, are usually seen from a human perspective as conditions to which deep-sea creatures have had to adapt. Through a different lens, those same elements enable life. They constitute conditions of livability. Pressure functions like a great muscle around each organism, providing containment of organs, holding bodies together. Moving up through the water column to escape mining plumes is no option for slow growing, slow-moving deep-sea creatures whose only escape is sideward across the ocean floor and bottom layer of seawater. Darkness, too, enables life in the benthos—creaturely existence depends on darkness to provide camouflage from predators, keep temperatures cool, and provide the conditions for bioluminescent communication between organisms.

3.1. Vulnerable Ocean Bodies and Material Flows

To build imaginaries that acknowledge the interconnectivity of vulnerable ocean bodies, we might trace how materials flow between and through embodied ocean beings in ways that render their living and long-dead kin, currents, and seafloor oozes ontologically inseparable. We could attune to how the ocean transitions through interconnections of nekton, shark, and krill navigating the water column; the dead bodies of whales, sunfish, and marine snow drifting to the seafloor; and the nutrient particles upwelled into the water column by the force of a deep benthic storm. Thinking with tuna, we can see how they exceed their economized value by coming into being and passing on the capacity for others to become, through flows of ocean materiality—briny matter moves through the fish's growing flesh and they flush and slough parts of themselves back into the solution in which they are suspended. As Stefanie Fishel and Susan Reid write, the tuna is "held by the ocean but at the same time, the ocean's omnipotent watery pressure pushes against its scaly, sleek body. Ocean floods the tuna's mouth becomes pressurized, and the tuna then squeezes separated oxygen through their billowing gills. Microbial particles flush from the tuna back into the ocean, along with organic matter and de-oxygenated water" (Celermajer et al. 2020). The fish is coconstitutive with the ocean—coming into being intra-actively with the ocean's minerals, chemicals, and organic materials.

Intimate and system-wide, material relationships expose the ocean to cumulative anthropogenic harms. It is these interconnections and vulnerabilities that developmentalism fails to notice. Under international fisheries regimes, described by Jennifer Telesca (2020: 3) as "exterminatory," unaccounted and undifferentiated beings are hauled from the sea and categorized under the vague term "bycatch." These beings came into existence within ecological communities. They are materially, ontologically, of the ocean. Along with the removal of the bodies of marine beings from the seas, the law facilitates industrial-scaled extraction of unseen relationships and material dependencies. In the nutrition-poor abyssal plains, historic and ongoing industrial fishing and whaling impacts deep-living beings, such as the jawless hagfish and bone-crunching zombie worms (Osedax) who are intimately connected to the nutritional flows of the water column, and acutely vulnerable to the loss of food "falls." Where once they provided abundant "bulk parcels" of organic enrichment for such detrivores (Higgs, Gates, and Jones 2014; Smith and Baco 2003), fewer carcasses of whales and other large bodies (falls) now drift to the seafloor. Changes in temperature and oxygen levels, associated with anthropogenic climate change, impact the rains of phytoplankton and microbes falling to the seafloor (Gittings et al. 2018). Lighter falls of marine snow mean fewer nutrients for sessile filter-feeders. Closer to shore, waste heat from a fossil fuel–reliant world nudges unlivable temperature changes into delicately balanced habitats, displacing vulnerable, embodied beings. Sea urchins on the move from tropical waters hitch rides on the warming East Australian Current that stretches southward due to climate change. Having reached Tasmania, the urchins are now devouring and disappearing underwater forests of the thirty-meter-tall giant kelp, *Macrosystis pyrifera* (van Sebille, Oliver, and Brown 2014). The arrival of new warm-water predators (Kelly 2011) is catching Tasmania's other cool-water residents unprepared, such as the yellowtail kingfish (Milman 2015), that are not suited to the warming water. As Erik van Sebille, Eric Oliver, and Jaci Brown (2014) note, these creatures can only move so far before they meet the edges of the continental shelf—this is the end of the line, as their next habitable shelf is about three thousand kilometers south in Antarctica.

Tracing material relations through the lens of vulnerability highlights how harm can be spatially and temporally distributed through the ocean. A sonic pulse blasted from an air gun reverberates outward across deep ocean clines and through the swim bladder of fish or the jaw fat of whales. Following the material flows of sonic energy reveals the spatialized nature of anthropogenic harm from the surface to the seafloor and across the water column. Additionally, vulnerability manifests spatiotemporally, as bivalves show us when they physically transition through distinct life stages. Bivalve larvae drift exposed in the water column for a period until they settle on the seabed to enter their next iteration of life as a slow growing sedentary adult. The same being occupies different bodily forms and ocean locations across their lifespan (Reid 2020). Vulnerabilities are imbricated spatiotemporally in cumulative layers of centuries of excessive fishing that intensify the impact of present fisheries regulations and corporate predation of marine animals.

The enduring potential of becoming vulnerable, which we all share (Fineman 2008: 12), imbues vulnerability with a dimension of futurity. At the same time, material streams fold us temporally forward, bound by knots of past consumptions and wastes. Ocean justice reaches back to understand the influence of past harms on present exposures and then casts forward to imagine how our material streams carry into distant, other futures. Alaimo (2016: 188) writes of these material, temporal distributions in the context of the violence of the Anthropocene's plastic streams, where marine beings are exposed to tides of "banal objects, intended for momentary human use [that] pollute for eternity" (130). Far from being remote actors, we are individually entangled with the vulnerabilities amplified and carried within these tides and the values that enable their proliferation. Alaimo captures these material flows in her figuration "transcorporeality"—the transit of matter across porous bodies and the ethical implications that these give rise to.

Ocean justice recognizes that the inherently transitional nature of the ocean is ruptured in kind and scale by the infections of anthropogenic waste from which extractive capital grossly disassociates itself. Virulent flows of plastic fog the ocean's interior and disrupt well-honed practices of trickery and trust that marine beings perform to catch their prey. Deep-living fish dangle bioluminescent lures to snare food, all the while hiding their powerful mouths and bodies in the dark-these body features and skills of trickery have been acquired over long-range processes of evolutions. Now the plastic contaminants of our commodity wastes perform a trickery that is rapidly, ubiquitously, and relentlessly transforming the ocean. Rather than delivering seasonal nutrients, current cargo is increasingly plastic (Parker 2020; Waller et al. 2017). Pushed deeper by surface currents, microplastics now pepper the ocean (Cressey 2016; Alaimo 2014), moving transcorporeally across gills and guts. Tiny plastic particulates are mistaken as food and ingested by fish, turtles, seabirds, and others—a slow violence (Nixon 2011) to which none is biologically adapted. As the unlikely distributors of capitalism's waste, marine beings send the plastic accumulated in their bellies further up the food chain—from predator to predator, including humans.[23] The ocean does not always disclose what they know—how is a fish or a seabird to recognize a cigarette butt as harmful? How are humans who eat fish protein to know that the "wild-caught" fish they ingest, or the bycatch that feeds farmed fish, are not already delivering doses of plastic?

Vulnerability is not exclusive to marine beings. Despite their planetary scale, the ocean's dynamic systems also have capacity to be affected by exposure to anthropogenic heat waste and biodiversity losses. Close to shore, the effects of climate change on the Great Barrier Reef (Readfearn 2020) are witnessed firsthand by experts,[24] media,[25] and citizen science initiatives (Small 2016). Deep in the hinterseas, anthropogenic heat waste forces material transitions in much less perceptible ways and is now altering current strength, reach, and turnover (IPCC 2019). The thermohaline current, which contributes to regulating planetary temperature distribution, is slowing (Nace 2020) because of changing salinity distribution arising from increasing polar ice melts and precipitation (Cai et al. 2005; Ridgeway 2007).

3.2. Vulnerability and Its Discursive Flows

Following Grear's (2015: 86–87) assertion that law is a "viral carrier of a hierarchy of being" and Alaimo's figuration of transcorporeality, I argue that vulnerability can be transmitted transcorporeally within the values and biases that stream through discursive bodies. The viral load of mastery's values and orientations to the planet can be detected across expanding extractivist frontiers through permissive regulatory frameworks, consumer decisions, and discards. Signature values of mastery, such as the dominance and instrumentalization of more-than-human others, are traceable across cultural bodies, government policy, and shareholder decisions and through the worlds unaccounted for in the profit margins of corporate humanity. These transcorporeal exchanges can have violent or benign consequences for planetary lives and potentials for cohabitation. From discursive body to discursive body, values harmful to ocean ecologies are translated by the machinery of extractivism into actual physical violences against the ocean and their marine beings.

Guided by the values of mastery, UNCLOS upholds a relationship of excessive exploitation against the seas. It presumes to bind us all to this relationship on the same terms. As the global framework convention, UNCLOS streams values of mastery through its legal frameworks to the International Seabed Authority, which passes them into regulations and implementation practices. These prescriptions influence other international and domestic ocean laws, economic policies, infrastructure investments, and corporate expansion strategies. The list is not exhaustive, and the networks are diffractive. Each institution, each policy document becomes a node for distributing harm through the values of mastery. UNCLOS's common heritage of mankind principle (CHM) is an example of mastery on steroids. It claims that the minerals of the international seabed area are there for the benefit of mankind[26] and distributes vulnerability through the values of mastery from which it was conceived.

Taylor writes that, given UNCLOS offers no firm definition of the common heritage principle, it might still be open to adding specific obligations (Taylor 1998: 297). Given these unsettled waters, the principle's key elements ought to be reimagined through an ocean justice lens. We might start with the principle's "heritage" element, which is commonly understood to mean wealth garnered from the extraction of seabed minerals. The doctrinal perspective invisibilizes the lively, complex realms of the seabed and deep ocean. Seatruthing the "heritage" element resists these denials by conceptually washing the deep ocean world back into the CHM principle. Seatruthing highlights that, as embodied beings, we interconnect with the seas through hydrological cycles and weather systems as well as mineral, chemical, protein, and nutrient kinships. Taylor argues that because CHM has "common interest" at its heart, it has potential to recognize the interdependence of earth systems and the interconnectedness of all human activities (278). Taking the above factors into account, the principle ought to recognize concepts of ecological subjectivity and humanity inclusively, and that our common interest includes both the more-than-human others that we co-become with and the oceanic realms of our prey. I propose that our common heritage is the relationships that suture us materially, temporally, and affectively with the ocean.

UNCLOS doesn't define the meaning of "mankind" in the "common heritage of mankind" principle. It does refer to "mankind as a whole"[27] though, which could reasonably be read as if a holistic concept of humankind were intended. However, ideologically and operationally the convention privileges corporate humanity. Corporations are the convention's paradigmatic legal subjects. It is collective corporate humankind that can partner with other companies and sponsoring states to operationalize mining; has the financial and technical resources to execute mining operations; can sell minerals on commodity markets; and rewards shareholders with profits. Without recognizing conditions of material embodiment, interconnectivity and vulnerability as well as concepts of ecological subjectivity, the principle limits how mining companies understand their interaction with the seabed and human and more-than-human marine others. As well, the intergenerational equity dimension folded into the principle assumes a fixed concept of both the ocean and humanity. The ocean of near and distant futures, and the material conditions faced by future humans, may well be difficult to recognize from today's vantage point.[28] Perhaps the consideration of an intergenerational equity element has more to do with maintaining status quo for the coming generations of corporate humanity who may be reliant on a continuity of mineral supplies and fish stock. Just as an ocean justice approach would reimagine "heritage" as our relationships with the ocean, it also reclaims the concept of "humanity" as one that includes all of us, not just the extractivists.

4. Institutions and Vulnerability

Fineman (2008: 19) argues that institutions ought to be responsive to, and responsible for, building the sustaining capacity of vulnerable subjects by providing adequate resources for resilience. One might expect such a "responsive state" (19) approach toward the ocean and the humans that depend on it from UNCLOS, given the convention's responsibilities for both extractive development and protection of the marine environment. However, as previously discussed, neither the convention nor the predatory liberal order that it privileges are epistemically competent to sufficiently recognize and respond to extractive development's impact on complex ocean vulnerabilities. In fact, as discussed at the beginning of this article, ecological conditions have deteriorated under its watch, undermining the convention's legitimacy.

Among the reasons that UNCLOS is unable to adequately protect the marine environment is the lack of clear guidelines for managing environmental impacts. There are no minimum standards for environmental impact statements, which are used by regulatory authorities to ensure environmental goals are being met (Wright et al. 2018: 35) and serve as tools for measuring vulnerability. Neither does UNCLOS explicitly reference the ecosystem approach (EA)—in part because the convention predates it. An EA is a strategy for the integrated management of environments, in which humans are placed within ecological contexts (De Lucia 2018: 105).[29] The use of EAs can be problematic, including when they overlap with ecosystem service concepts that represent ecological worlds within economic terms. They also risk biopolitical slippage between care for ecosystems and their subjugation, as Vito De Lucia notes (113). It will be difficult to realize the promise of EAs and adequately recalibrate toward more ecological relationships with the sea while governance regimes continue to deny material embodiment and interconnectivity. Nevertheless, the presence of EAs within governance or policy instruments signals potential shifts toward less anthropocentric legal frameworks (18) and recognition of ecological vulnerability (Weißhuhn, Muller, and Wiggering 2018).

There are already many examples of ecosystem approaches within marine environmental law. Under the framework of UNCLOS, a new international legally binding instrument is being negotiated for the protection of marine biodiversity in the high seas areas beyond national jurisdiction.[30] While the Draft Text remains fundamentally anthropocentric and with many provisions still unsettled, it makes novel departures from UNCLOS.[31] It hints at the potential for an international governance regime that could be more ecologically sensitive to marine vulnerability. For example, article 5 includes "(f) An ecosystem approach," along with other ecologically oriented elements: "(e) The precautionary [principle] [approach]; [(g) An integrated approach;] (h) An approach that builds ecosystem resilience to the adverse effects of climate change and ocean acidification and restores ecosystem integrity."[32] That said, the more ecologically oriented provisions are included as "General [principles] [and] [approaches]," not obligations. As De Lucia argues, at this stage it would appear that implementation risks being inconsistent at best, given the general ambiguity around the EA concept and the Draft Text's lack of clear guidance on how nations ought to implement it in their management framework (De Lucia 2019).[33]

The developing stages of the Draft Text demonstrate how discursive elements and knowledge environments function ecologically to produce vulnerability. It portends what possible future relationship with the ocean will be countenanced by nations and their corporate sponsors, and what harms will be tolerated. Moreover, the dynamics of inclusion and exclusion are present not only in substance but also in process. The ethos and hierarchies of mastery prevail to determine who contributes to the Draft Text, which voices are heard, which are absent, and which worldviews get recognized. It is a pattern of treaty development that large and well-resourced nations and privileged humans dominate proceedings to the exclusion of smaller nations and more vulnerable subjects within those nations. With limited access to funding and international legal expertise, smaller nations are restricted in their ability to contribute submissions or to attend the Intergovernmental Conferences where substantive agreement matters are negotiated (Tessnow-von Wysocki 2019). The treaty-making proceedings include detailed negotiations over definitions and implications of intellectual property and financial benefit sharing of genetic resources. By contrast, the Draft Text generalizes the marine realms captured by the high seas jurisdiction, and which make up about 95 percent of the Earth's occupied habitat, as "the ocean." Absent are the material, social, and dynamic relationships and more-than-human constituents that will be affected by this emerging, new "architecture of exploitation" (Reid 2022: 73). Neither will the existing, exterminatory fisheries regimes likely be impinged on by the new agreement. The Draft Text and UNCLOS exemplify the political and onto-epistemic challenges of current ocean governance approaches. While concepts such as the ecosystem approach and precautionary principle create discursive spaces within international law for the countenance of specific marine protections, the underlying levers of mastery and anthropocentricism patrol the limits of those protections, and who and what counts.

Reflecting on the institutional components of vulnerability is important because protection of the seas, particularly in remote areas, is well beyond the capacity of individuals. It is a responsibility that necessarily falls on nation-states acting alone and collectively, or under the framework of international legal instruments and cooperation. Responsive state obligations raise another less obvious issue of vulnerability, which is that people unfamiliar with the operations and limitations of international law might believe that current governance regimes are adequately protecting the ocean against harm. How many among them would know that UNCLOS and the Seabed Authority together operate as a powerful architecture of exploitation that legitimates the extractive development of the international seabed area? Regimes such as UNCLOS are entrusted to protect the ocean for us all, not just for corporations or the global economic order. Their existence provides misleading reassurances, pacifying concerns. This is part of the ideological function of law—it holds a place in the cultural imagination as the protector of norms and the vulnerable but in fact is frequently the agent of domination.

Reviewing the Course

International law controls the dominant relationship with the ocean. This essay attributes the declining conditions of the ocean to the exploitative nature of that relationship. To reimagine how a less harmful relationship could take form, we need to think with the ocean, outside the constraints of law's Enlightenment imaginary. My ocean justice proposal contributes to that process but does not presume to offer analytical certainty or sure-fix solutions. Ocean justice commences with a desire to cohabitate well with the seas and draws on multiple voices and knowledges to build the conditions of possibility for that to happen. The relational shift discussed in the context of the common heritage of mankind principle locates where the transformation might begin within the foundations of law—that is, by reimagining "common heritage" not as the wealth that can be accumulated by exploiting the ocean but as the relationships that suture us to the ocean, including with affection and respect. The "humanity" to be reclaimed in the principle is all of us, as ecological subjects whose well-being is bound inextricably with the seas.

Acknowledgments

This article was written on the unceded, stolen lands of Gadigal and Yugambeh peoples.

Notes

1. Halpern and coauthors found that 59 percent of the ocean is experiencing cumulative impacts due to human activities and climate change (Halpern et al. 2019). Declining marine ecological conditions are widely reported: fish extraction (Pauly et al. 1998; Pauly 2010; Telesca 2020; Probyn 2016); pollution (Gross 2015; Cressey 2016; Arias and Botte 2020; Probyn 2018); heat and carbon (IPCC 2019); sonic impacts (Mitchell 2009; Roberts 2007; Duarte et al. 2021); and more broadly (Ramirez-Llodra et al. 2011; IPCC 2019; United Nations 2021).
2. The United Nations Convention on the Law of the Sea (adopted December 10, 1982, entered into force November 16, 1994), 1833 UNTS 397 (hereafter, UNCLOS).
3. Despite increasing scientific attention and cumulative knowledge, just over 10 percent of ocean lives have been described (Mora et al. 2011: 2), and as of 2017 only 6 percent of the ocean had been mapped (Trethewey 2020).
4. I note Amanda Mosborg (2020) uses the term *ocean justice* in the context of human rights and social justice.
5. I deliberately use the gender-neutral pronoun *they/their/them* to recognize the ocean as not an object but as a gender-neutral, collective planetary entity.
6. For more on seabed mining and the complicity of governance regimes, see Reid 2020, 2022; Ranganathan 2019.

7. See also Jue's (2020: 4–6) argument that immersing theory into the ocean's milieu can give rise to generative "conceptual displacements."
8. I use *prosthetics* in the sense that Donna Haraway (1988: 583) intended the term to describe technologies and devices through which knowledge is mediated.
9. See, for example, Bolivia: Ley de Derechos de la Madre Tierra 2010 (Bolivia), article 3; India: Mohd Salim v. State of Uttarakhand and others, WPPIL 126/2014 (High Court of Uttarakhand) 2017 [19]; and New Zealand: Te Awa Tupua (Whanganui River Claims Settlement) Act 2017, No. 7, Public Act. For discussions about earth jurisprudence and rights of nature, see Burdon 2011; Murray 2015; Rogers and Maloney 2017.
10. For discussion about ecological law and mining, see Sbert 2020.
11. For discussions concerning Indigenous kinship relations, stewardship models, and the impacts of colonization, see LaDuke 2016; Country et al. 2016; Ingersoll 2016; Todd 2014, 2017a, 2017b; Gómez-Barris 2017; Marshall 2017; Taylor and Parsons 2021.
12. Ecological integrity is a familiar "common conceptual denominator" within international environmental law (Bosselmann and Rakhyun 2015: 194).
13. Along with other gendered, racialized, and ableist privileging.
14. The common heritage of mankind principle applies to the seabed area beyond national jurisdiction (UNCLOS Preamble para 6).
15. In their discussion of ecofeminist-informed ecological economics, Ruder and Sanniti describe "predatory ontologies" as "the structural externalisation that enables enactments of injustices by predatory global corporate neo-liberalism" (Ruder and Sanniti 2019).
16. ISBA/25/C/WP.1. (Draft Exploitation Regulations).
17. ISBA/25/C/WP.1., reg. 2 (e).
18. ISBA/25/C/WP.1., regs. 2 (e)(ii) and 44 (a).
19. Hatched at the 1992 Conferences on Environment and Development in Rio de Janeiro and instituted in Principle 15 of the RIO Convention: "In order to protect the environment, the precautionary approach shall be widely applied by States according to their capabilities. Where there are threats of serious or irreversible damage, lack of full scientific certainty shall not be used as a reason for postponing cost-effective measures to prevent environmental degradation."
20. Louisa Casson and coauthors (2020) investigated the operations of the Seabed Authority and found evidence of systemic, regulatory capture of the authority by corporations (Casson 2020).
21. The Clarion Clipperton Zone (CCZ) is an international seabed area located in the North Pacific Ocean, where the first deep seabed mining activities are likely to commence. For more on potential ecological risks, see Macheriotou et al. 2020.
22. See also Jue's (2020) account of the ocean as media and Peters and Steinberg's theorization of wet ontology (Peters and Steinberg 2019).
23. Humans now ingest fish and salt laced with plastic; see Thompson 2018.
24. Such as the Great Barrier Reef Marine Park Authority (n.d.), which monitors and researches the impacts of climate change on the reef.
25. For example, Atlantic Productions, "David Attenborough's Great Barrier Reef."
26. UNCLOS art 136; Preamble para 6. Note that this article uses the gender-neutral *humankind*.
27. UNCLOS Preamble para 5; art 137.
28. Davis (2015: 239) writes that the concept of intergenerational equity is fraught within international law, as it projects a "fantasy of continuance" in which future children are insulated from environmental change.
29. It is not all smooth seas for EAs. In her posthumanist critique of international ocean law, Alice Ollino (2019) argues that the inclusion of EAs in fisheries governance unsettles the anthropocentricism of law and indicates an emerging ecocentric approach. Jennifer Telesca's (2020) ethnographic account of regional fisheries management reveals anthropocentricism prevails in the dominant, statistically based resource management models. See also De Lucia 2018: 105–6.
30. UN, Revised Draft Text of an Agreement under the United Nations Convention on the Law of the Sea on the Conservation and Sustainable Use of Marine Biological Diversity of Areas

beyond National Jurisdiction. UN Doc. A/CONF.232/2019/6 (hereafter, Draft Text).

31. This includes recognition of Indigenous traditional knowledges. The Draft Text provides that parties shall be guided by integrating the use of relevant traditional knowledge of Indigenous peoples and local communities (article 5), including in identifying marine areas for protection (article 6), and consultation toward their establishment (article 17). The conflicting textual proposals submitted by national delegations in response to the Draft Text demonstrates the contested nature of these and other provisions: https://www.un.org/bbnj/sites/www.un.org.bbnj/files/a_conf232_2022_inf1_textualproposalscompilation_articlebyarticle15april2020_rev.pdf.
32. Draft Text, article 5.
33. See also De Lucia 2018: 113; 2019, 12.

References

Alaimo, Stacy. 2014. "Oceanic Origins Plastic Activism and New Materialism at Sea." In *Material Ecocriticism*, 186–203. Bloomington: Indiana University Press.

Alaimo, Stacy. 2016. *Exposed: Environmental Politics and Pleasures in Posthuman Times*. Minneapolis: University of Minnesota Press.

Alaimo, Stacy. 2019. "Wanting All the Species to Be: Extinction, Environmental Visions, and Intimate Aesthetics." *Australian Feminist Studies*, no. 102: 398–412. https://doi.org/10.1080/08164649.2019.1698284.

Arias, Andrés Hugo, and Sandra Elizabeth Botte. 2020. *Coastal and Deep Ocean Pollution*. Boca Raton: CRC.

Atlantic Productions. n.d. "David Attenborough's Great Barrier Reef." https://attenboroughsreef.com/ (accessed October 26, 2022).

Bosselmann, Klaus, and Kim E. Rakhyun. 2015. "Operationalizing Sustainable Development: Ecological Integrity as a Grundnorm of International Law." *Review of European, Comparative, and International Environmental Law (RECIEL)* 24, no. 2: 194–208. https://doi.org/10.4324/9781315553396.

Burdon, Peter, ed. 2011. *Exploring Wild Law: The Philosophy of Earth Jurisprudence*. Kent Town, SA: Wakefield.

Cai, W., G. Shi, T. Cowan, D. Bi, and J. Ribbe. 2005. "The Response of the Southern Annular Mode, the East Australian Current, and the Southern Mid-Latitude Ocean Circulation to Global Warming." *Geophysical Research Letters* 32, no. 23. https://doi.org/10.1029/2005GL024701.

Casson, Louisa, et al. 2020. "Deep Trouble: The Murky World of the Deep Sea Mining Industry." Amsterdam, Netherlands: Greenpeace International. https://www.greenpeace.org/international/publication/45835/deep-sea-mining-exploitation/.

Celermajer, Danielle. 2018. *The Prevention of Torture: An Ecological Approach*. Cambridge: Cambridge University Press.

Celermajer, Danielle, Sria Chatterjee, Alasdair Cochrane, Stefanie Fishel, Astrida Neimanis, Anne O'Brien, Susan Reid, Kritika Srinivasan, David Schlosberg, and Anik Waldow. 2020. "Justice through a Multispecies Lens." *Contemporary Political Theory* 19, no. 3: 475–512. https://doi.org/10.1057/s41296-020-00386-5.

Country, Bawaka, Sarah Wright, Sandie Suchet-Pearson, Kate Lloyd, Laklak Burarrwanga, Ritjilili Ganambarr, Merrkiyawuy Ganambarr-Stubbs, Banbapuy Ganambarr, Djawundil Maymuru, and Jill Sweeney. 2016. "Co-becoming Bawaka: Towards a Relational Understanding of Place/Space." *Progress in Human Geography* 40, no. 4: 455–75. https://doi.org/10.1177/0309132515589437.

Cressey, Daniel. 2016. "Bottles, Bags, Ropes and Toothbrushes: The Struggle to Track Ocean Plastics." *Nature* 536: 263–65. https://www.nature.com/articles/536263a.

Davis, Heather. 2015. "Toxic Progeny: The Plastisphere and Other Queer Futures." *Philosophia* 5, no. 2: 231–50.

De Lucia, Vito. 2018. "A Critical Interrogation of the Relation between the Ecosystem Approach and Ecosystem Services." *Review of European, Comparative, and International Environmental Law* 27, no. 2: 104–14. https://doi.org/10.1111/reel.12227.

De Lucia, Vito. 2019. "The Ecosystem Approach and the BBNJ Negotiations." *2 Nordic Journal of Environmental Law*, July 17. https://doi.org/10.2139/ssrn.3420988.

Duarte, Carlos M., et al. 2021. "The Soundscape of the Anthropocene Ocean." *Science* 371. https://doi.org/10.1126/science.aba4658.

Fineman, Martha Albertson. 2008. "The Vulnerable Subject: Anchoring Equality in the Human Condition." *Yale Journal of Law and Feminism* 20, no. 1: 1–24.

Fineman, Martha, and Anna Grear, eds. 2013. *Vulnerability: Reflections on a New Ethical Foundation for Law and Politics*. London: Ashgate.

Fraser, Nancy. 2003. *Redistribution or Recognition? A Political-Philosophical Exchange*. London: Verso.

Gittings, John A., Dionysios E. Raitsos, George Krokos, and Ibrahim Hoteit. 2018. "Impacts of Warming on Phytoplankton Abundance and Phenology in a Typical Tropical Marine Ecosystem." *Scientific Reports* 8, no. 1: 2240. https://doi.org/10.1038/s41598-018-20560-5.

Gómez-Barris, Macarena. 2017. *The Extractive Zone: Social Ecologies and Decolonial Perspectives*. Durham, NC: Duke University Press. https://doi.org/10.1215/9780822372561.

Grear, Anna. 2015. "The Closures of Legal Subjectivity: Why Examining 'Law's Person' Is Critical to an Understanding of Injustice in an Age of Climate Crisis." In *Research Handbook on Human Rights and the Environment*, edited by Anna Grear and Louis J. Kotzé, 79–101. Cheltenham, UK: Edward Elgar.

Great Barrier Reef Marine Park Authority. n.d. "Reef Health." https://www2.gbrmpa.gov.au/learn/reef-health.

Gross, Michael. 2015. "Oceans of Plastic Waste." *Current Biology* 25, no. 3: R93–R96. https://doi.org/10.1016/j.cub.2015.01.038.

Halpern, Benjamin S., Melanie Frazier, Jamie Afflerbach, Jujia S. Lowndes, Fiorenza Micheli, Casey O'Hara, Courtney Scarborough, and Kimberly A. Selkoe. 2019. "Recent Pace of Change in Human Impact on the World's Ocean." *Scientific Reports* 9: 11609.

Haraway, Donna J. 1988. "Situated Knowledges: The Science Question in Feminism and the Privilege of Partial Perspective." *Feminist Studies* 14, no. 3: 575–99.

Haraway, Donna J. 2016. *Staying with the Trouble: Making Kin in the Chthulucene*. Durham. NC: Duke University Press.

Higgs, Nicholas D., Andrew R. Gates, and Daniel O. B. Jones. 2014. "Fish Food in the Deep Sea: Revisiting the Role of Large Food-Falls." *PLoS One* 9, no. 5: e96016. https://doi.org/10.1371/journal.pone.0096016.

Ingersoll, Karin. E. 2016. *Waves of Knowing: A Seascape Epistemology*. Durham, NC: Duke University Press.

IPCC (Intergovernmental Panel on Climate Change). 2019. *Special Report on the Ocean and Cryosphere in a Changing Climate*. https://www.ipcc.ch/srocc/.

Jackson, Moana. 1993. "Indigenous Law and the Sea." In *Freedom for the Seas in the Twenty-First Century: Ocean Governance and Environmental Harmony*, edited by Jon M. Van Dyke, Durwood Zaelke, and Grant Hewison, 41–48. Washington, DC: Island.

Jue, Melody. 2020. *Wild Blue Media: Thinking through Seawater*. Durham, NC: Duke University Press.

Kelly, Kathryn, dir. 2011. *The Inertia Trap: Climate Change and the Oceans*. Video file. Mitchell, ACT: Ronin Films.

Kotzé, Louis. 2019. "The Anthropocene, Earth System Vulnerability and Socio-Ecological Injustice in an Age of Human Rights." *Journal of Human Rights and the Environment* 10, no. 1: 62–85. https://doi.org/10.4337/jhre.2019.01.04.

LaDuke, Winona. 2016. *All Our Relations: Native Struggles for Land and Life*. Chicago: Haymarket.

Lather, Patti. 1993. "Fertile Obsession: Validity after Poststructuralism." *The Sociological Quarterly* 34, no. 4: 673–93. http://www.jstor.org/stable/4121374.

Lorde, Audre. 1984. *Sister Outsider: Essays and Speeches*. Berkeley, CA: Crossing.

Macheriotou, Lara, Annelien Rigaux, Sofie Derycke, and Ann Vanreusel. 2020. "Phylogenetic Clustering and Rarity Imply Risk of Local Species Extinction in Prospective Deep-Sea Mining Areas of the Clarion-Clipperton Fracture Zone." *Proceedings of the Royal Society B: Biological Sciences*, no. 1924: 20192666. https://doi.org/10.1098/rspb.2019.2666.

Marshall, Virginia. 2017. *Overturning Aqua Nullius: Securing Aboriginal Water Rights*. Sydney: Aboriginal Studies Press.

Milman, Oliver. 2015. "Australian Fish Moving South as Climate Changes, Say Researchers." *Guardian*, January 29, 2015. http://www.theguardian.com/environment/2015/jan/29/australian-fish-moving-south-as-climate-changes-say-researchers.

Mitchell, Alanna. 2009. *Seasick: The Hidden Ecological Crisis of the Global Ocean*. Chicago: University of Chicago Press.

Mora, Camilo, Derek P. Tittensor, Sina Adl, Alastair G. Simpson, and Boris Worm. 2011. "How Many Species Are There on Earth and in the Ocean?" *PLoS Biology* 9, no. 8: e1001127.

Mosborg, Amanda. 2020. "Human Rights on the Ocean // Ocean Justice." *The Ocean Foundation* (blog), September 11. https://oceanfdn.org/human-rights-on-the-ocean-ocean-justice/.

Murray, Jamie. 2015. "Earth Jurisprudence, Wild Law, Emergent Law: The Emerging Field of Ecology and Law." Pt. 2. *Liverpool Law Review* 36, no. 2: 105–22. https://doi.org/10.1007/s10991-015-9170-y.

Nace, Trevor. 2020. "Global Ocean Circulation Keeps Slowing Down: Here's What It Means." *Forbes*, November 27. https://www.forbes.com/sites/trevornace/2018/11/27/global-ocean-circulation-keeps-slowing-down-heres-what-it-means/.

Naffine, Ngaire. 2003. "Who Are Law's Persons? From Cheshire Cats to Responsible Subjects." *Modern Law Review* 66, no. 3: 346–67. https://doi.org/10.1111/1468-2230.6603002.

Neimanis, Astrida. 2014. "Alongside the Right to Water, a Posthumanist Feminist Imaginary." *Journal of Human Rights and the Environment* 5, no. 1: 5–24.

Nixon, Rob. 2011. *Slow Violence and the Environmentalism of the Poor*. Cambridge, MA: Harvard University Press.

Ollino, Alice. 2019. "Feminism, Nature, and the Post-human: Toward a Critical Analysis of the International Law of the Sea Governing Marine Living Resources Management." In *Gender and the Law of the Sea*, edited by Irini Papanicolopulu, 204–28. https://doi.org/10.1163/9789004375178_011.

Parker, Laura. 2020. "Microplastics Have Moved into Virtually Every Crevice on Earth." *National Geographic*, August 7, 2020. https://www.nationalgeographic.com/science/2020/08/microplastics-in-virtually-every-crevice-on-earth/.

Pauly, Daniel. 2010. *Five Easy Pieces: The Impact of Fisheries on Marine Ecosystems*. Washington, DC: Island.

Pauly, Daniel, Villy Christensen, Johanne Dalsgaard, Rainer Froese, and Francisco Torres Jr. 1998. "Fishing Down Marine Food Webs." *Science* 279: 860–63. https://doi.org/10.1126/science.279.5352.860.

Peters, Kimberley, and Philip Steinberg. 2019. "The Ocean in Excess: Towards a More-than-Wet Ontology." *Dialogues in Human Geography* 9, no. 3: 293–307. https://doi.org/10.1177/2043820619872886.

Philippopoulos-Mihalopoulos, Andreas. 2011. "Towards a Critical Environmental Law." In *Law and Ecology: New Environmental Foundations*, edited by Andreas Philippopoulos-Mihalopoulos, 18–38. New York: Routledge.

Philippopoulos-Mihalopoulos, Andreas. 2017. "Critical Environmental Law as Method in the Anthropocene." In *Research Methods in Environmental Law*, edited by Andreas Philippopoulos-Mihalopoulos and Victoria Brooks, 131–55. http://www.elgaronline.com/view]/edcoll/9781784712563/9781784712563.00012.xml.

Plumwood, Val. 2009. "Nature in the Active Voice." *Australian Humanities Review*, no. 46: 113–29. https://doi.org/10.22459/AHR.46.2009.10.

Probyn, Elspeth. 2016. *Eating the Ocean*. Durham: Duke University Press.

Probyn, Elspeth. 2018. "The Ocean Returns: Mapping a Mercurial Anthropocene." *Social Science Information* 57, no. 3: 386–402. https://doi.org/10.1177/0539018418792402.

Puig de la Bellacasa, Maria. 2012. "'Nothing Comes without Its World': Thinking with Care." *Sociological Review* 60, no. 2: 197–216. https://doi.org/10.1111/j.1467-954X.2012.02070.x.

Ramirez-Llodra, E., et al. 2011. "Man and the Last Great Wilderness: Human Impact on the Deep Sea." *PloS One* 6, no. 8: e22588. https://doi.org/10.1371/journal.pone.0022588.

Ranganathan, Surabhi. 2019. "Ocean Floor Grab: International Law and the Making of an Extractive Imaginary." *European Journal of International Law* 30, no. 2: 573–600. http://www.ejil.org/archive.php?issue=148.

Readfearn, Graham. 2020. "Great Barrier Reef's Third Mass Bleaching in Five Years the Most Widespread Yet." *Guardian*, April 6. https://www.theguardian.com/environment/2020/apr/07/great-barrier-reefs-third-mass-bleaching-in-five-years-the-most-widespread-ever.

Reid, Susan. 2020. "Solwara 1 and the Sessile Ones." In *Blue Legalities: The Life and Laws of the Sea*, edited by Irus Braverman and Elizabeth R. Johnson, 25–44. Durham, NC: Duke University Press. https://doi.org/10.1215/9781478007289-002.

Reid, Susan. 2022. "Imagining Justice with the Abyssal Ocean." In *Laws of the Sea: Interdisciplinary Currents*, edited by Irus Braverman, 69–92. New York: Routledge.

Reuters. 2012. "RPT-NZ Refining Shareholders Approve Expansion Project." April 27. https://www.reuters.com/article/nzrefining/rpt-nz-refining-shareholders-approve-expansion-project-idUSW9E8E701F20120427.

Ridgway, K. R. 2007. "Long-Term Trend and Decadal Variability of the Southward Penetration of the East Australian Current." *Geophysical Research Letters* 34, no. 13. https://doi.org/10.1029/2007GL030393.

Roberts, Callum. 2007. *The Unnatural History of the Sea*. Washington, DC: Island/Shearwater.

Rogers, Nicole, and Michelle Maloney. 2017. *Law as if Earth Really Mattered: The Wild Law Judgment Project*. New York: Routledge. https://doi.org/10.4324/9781315618319.

Ruder, Sara-Louise, and Sophia Rose Sanniti. 2019. "Transcending the Learned Ignorance of Predatory Ontologies: A Research Agenda for an Ecofeminist-Informed Ecological Economics." *Sustainability* 11, no. 5: 1479. https://doi.org/10.3390/su11051479.

Satz, Ani B. 2009. "Animals as Vulnerable Subjects: Beyond Interest-Convergence, Hierarchy, and Property." *Animal Law* 16, no. 1: 122.

Sbert, Carla. 2020. *The Lens of Ecological Law: A Look at Mining*. Northampton, UK: Edward Elgar. https://doi.org/10.4337/9781839102134.

Shotwell, Alexis. 2016. *Against Purity: Living Ethically in Compromised Times*. Minneapolis: University of Minnesota Press.

Singh, Juliette. 2017. *Unthinking Mastery: Dehumanism and Decolonial Entanglements*. Durham, NC: Duke University Press.

Small, Stephanie. 2016. "Great Barrier Reef: Citizen Scientists Lend a Hand to Document Coral Bleaching Impact." *ABC News*, October 22. https://www.abc.net.au/news/2016-10-22/great-barrier-reef-citizen-scientists-help-coral-bleaching/7957414.

Smith, Craig R., and Amy R. Baco. 2003. "Ecology of Whale Falls at the Deep-Sea Floor." In *Oceanography and Marine Biology: An Annual Review*, 41: 311–54. London: Taylor and Francis.

Steinberg, Philip, and Kimberley Peters. 2015. "Wet Ontologies, Fluid Spaces: Giving Depth to Volume through Oceanic Thinking." *Environment and Planning D: Society and Space* 33, no. 2: 247–64. https://doi.org/10.1068/d14148p.

Taylor, Lara, and Meg Parsons. 2021. "Why Indigenous Knowledge Should Be an Essential Part of How We Govern the World's Oceans." *Conversation*, June 8. http://theconversation.com/why-indigenous-knowledge-should-be-an-essential-part-of-how-we-govern-the-worlds-oceans-161649.

Taylor, Prue. 1998. *An Ecological Approach to International Law: Responding to Challenges of Climate Change*. London: Routledge. https://doi.org/10.4324/9780203031186.

Telesca, Jennifer E. 2020. *Red Gold: The Managed Extinction of the Giant Bluefin Tuna*. Minneapolis: University of Minnesota Press.

Tessnow-von Wysocki, Ina. 2019. "Slow Progress in the Third BBNJ Meeting: Negotiations Are Moving—but Sideways." *Maripoldata*, September 6. https://www.maripoldata.eu/slow-progress-in-the-third-bbnj-meeting-negotiations-are-moving-but-sideways/.

Thompson, Andrea. 2018. "From Fish to Humans, a Microplastic Invasion May Be Taking a Toll. *Scientific American*, September 4. https://www.scientificamerican.com/article/from-fish-to-humans-a-microplastic-invasion-may-be-taking-a-toll/.

Trethewey, Laura. 2020. "Earth's Final Frontier: The Global Race to Map the Entire Ocean Floor." *Guardian*, June 30. https://www.theguardian.com/environment/2020/jun/30/earths-final-frontier-the-global-race-to-map-the-entire-ocean-floor.

Todd, Zoe. 2014. "Fish Pluralities: Human-Animal Relations and Sites of Engagement in Paulatuuq, Arctic Canada." *Etudes Inuit* 38, nos. 1–2: 217–38. https://doi.org/10.7202/1028861ar.

Todd, Zoe. 2017a. "Fish, Kin and Hope: Tending to Water Violations in Amiskwaciwâskahikan and Treaty Six Territory." *Afterall: A Journal of Art, Context and Enquiry* 43: 102–7. https://doi.org/10.1086/692559.

Todd, Zoe. 2017b. "Protecting Life Below Water: Tending to Relationality and Expanding Oceanic Consciousness Beyond Coastal Zones." *Journal of the American Anthropological Association*. https://www.americananthropologist.org/deprovincializing-development-series/protecting-life-below-water.

United Nations. 2021. *The Second World Ocean Assessment*. 2 vols. https://www.un.org/regularprocess/sites/www.un.org.regularprocess/files/2011859-e-woa-ii-vol-i.pdf; https://www.un.org/regularprocess/sites/www.un.org.regularprocess/files/2011859-e-woa-ii-vol-ii.pdf.

van Sebille, Erik, Eric Oliver, and Jaci Brown. 2014. "Can You Surf the East Australian Current, Finding Nemo-Style?" *Conversation*, June 5. http://theconversation.com/can-you-surf-the-east-australian-current-finding-nemo-style-27392.

Waller, Catherine L., Huw J. Griffiths, Claire M. Waluda, Sally E. Thorpe, Iván Loaiza, Bernabé Moreno, Cesar O. Pacherres, and Kevin A. Hughes. 2017. "Microplastics in the Antarctic Marine System: An Emerging Area of Research." *Science of The Total Environment* 598: 220–27. https://doi.org/10.1016/j.scitotenv.2017.03.283.

Weißhuhn, Peter, Felix Muller, and Hubert Wiggering. 2018. "Ecosystem Vulnerability Review: Proposal of an Interdisciplinary Ecosystem Assessment Approach. (Report)." *Environmental Management* 61, no. 6: 904–15. https://doi.org/10.1007/s00267-018-1023-8.

Wright, G., J. Rochette, K. Gerdje, and I. Seeger. 2018. *The Long and Winding Road: Negotiating a Treaty for the Conservation and Sustainable Use of Marine Biodiversity in Areas beyond National Jurisdiction*. IDDRI Studies 08/18. https://www.iddri.org/sites/default/files/PDF/Publications/Catalogue%20Iddri/Etude/201808-Study_HauteMer-long%20and%20winding%20road.pdf.

Yusoff, Kathryn. 2012. "Aesthetics of Loss: Biodiversity, Banal Violence, and Biotic Subjects: Aesthetics of Loss." *Transactions of the Institute of British Geographers* 37, no. 4: 578–92. https://doi.org/10.1111/j.1475-5661.2011.00486.x.

Susan Reid is a creative researcher and cultural theorist, artist, writer, and lawyer whose research interests include multibeing justice and ocean relationalities. She is a research affiliate with Sydney University's ARC funded project Extracting the Ocean and a member of the Sydney Environment Institute. Susan is of mixed Celtic/Anglo/European settler ancestry, living nomadically on unceded Gadigal and Yugambeh lands.

TREE STORIES

The Embodied History of Trees and Environmental Ethics

Dalia Nassar and Margaret Barbour

Abstract **This article develops the notion of the "embodied history of trees" and articulates its conceptual and ethical implications. It demonstrates how trees literally embody their environment in their very structure and argues that trees express their environments in the deepest, most responsive, and most immediate way. The article then moves to consider how trees fundamentally shape their environment, showing that just as trees are expressions of their contexts, so their contexts are expressions of the trees. By highlighting the deep reciprocity between trees and their environments, the article raises crucial questions about the usual modes of conceptualizing the relation between organism and environment, and points to the ways in which environmental ethics remains largely wedded to these problematic conceptualizations. It concludes by developing environmental ethical concepts in light of the embodied history of trees, noting how these concepts challenge assumptions within mainstream environmental ethics, while extending the insights of deep ecology, ecofeminism, and Indigenous relational ethics in illuminating ways.**

Keywords **trees, plants, embodied history, entanglement, environment, ethics, representation**

Kia ora e te Rimu—*Hello Rimu My Friend*

I'm standing at your base, and my outstretched arms reach less than halfway around you. I give you a half-hug as I look upward to your moss-furred, sky-reaching branches, hoping that my feet aren't hurting too many of your delicate surface

Cultural Politics, Volume 19, Issue 1, © 2023 Duke University Press
DOI: 10.1215/17432197-10232530

roots. As my fingertips reach through the epiphytic wrap to touch your rough bark, I smell your sharp, astringent top note among the deep humus base of the rainforest, and I wonder: will I ever get to know you?

From looking at your annual growth rings in a core I extracted last year, I know that you are about four hundred years old. I know from the velocity probes I drilled into you that your xylem sap moves fastest at the middle of bright sunny days and slows to zero in the dark. This mirrors the daily rhythm of the stomatal pores on your midcanopy leaves that I measured over long, sleepy summer days after hauling gas exchange equipment more than twenty meters up the canopy access tower. Those same hot summer days, with a big high-pressure system squatting over the Tasman Sea, turned your moss-enfolded branches into warm, furry teddy bear arms. In my mind's eye I can see the clever alchemy of your photosynthetic biochemistry, and I can track that you've used the stored energy to build new leaves, new roots, and a sturdier stem.

Maybe I'm starting to know a little of your world, Rimu.[1] Simply by standing with you, I have become physically entangled with you and your life. My interest in you and the instruments I use to measure your function and behavior entangle us even further. Might this be—to speak with Karen Barad—the beginning of knowledge? As Barad (2007: 185) puts it, "We don't obtain knowledge by standing outside the world; we know because we are of the world."

But another look at your pendulous leaves stops me short. Can I ever be of *your* world? It feels like an unbridgeable chasm separates your world and mine. Still, I attempt this bridge, with the inkling that getting to know you is crucially important for your sake—and also for mine.

1. Introduction: Why Trees?

Why is getting to know Rimu, an evergreen coniferous tree species (*Dacrydium cupressinum* Sol. ex Lamb.) endemic to New Zealand, so important? Why should we turn our attention to a being that—like all trees—does not have a clear "face" or an evident "voice," that towers above us, alive for hundreds of years and resilient in ways we cannot imagine?

Trees are, from one perspective, the most important life-form on earth. Every schoolchild learns the basic facts: trees provide us with sustenance, and their photosynthetic activity creates an atmosphere that enables our survival. Or, as anthropologist Natasha Myers (2020) colorfully puts it, trees "breathed us into being." Without trees (and marine phytoplankton), the earth would be uninhabitable, and with their rising rates of death and extinction, the earth might very well become uninhabitable.

Trees also populate our imagination. Many of us first become interested in trees through legends and traditional stories. Forests loom large in European folklore, whether as dark and fearful places, magical wonderlands, or safe havens. They are almost always figured as sites of human transformation. In Australian Indigenous cultures, trees are regarded as community members, both to be protected and to provide protection (Troy 2019). Within te ao Māori (the culture of New Zealand's first people), the large forest tree Tōtara (*Podocarpus totara*) is the first born of the god Tāne, with humans as younger siblings. Hence, the *whakapapa* (kinship lines, or genealogy) of humans includes trees and other plants.

However, despite their manifold significance, trees—and plants more generally—have been largely relegated to the sidelines of Western philosophical, environmental, and ethical debates, leading some to speak of "plant blindness" (Wandersee and Schussler 1999) and others to argue that in contemporary theory plants have a "ghost-like presence" (Jones and Cloke 2002). While plants have received some attention in the last decade (Hall 2011; Marder 2013; Head et al. 2014; Coccia 2019), the majority of scholars within environmental ethics and humanities have (until very recently) focused on animal rights, animal suffering, and our moral responsibility to animals (e.g., Singer 1975; Regan 1983; Nussbaum 2006; Haraway 2008; Korsgaard 2018). Similarly, trees are rarely considered in theoretical discussions of living beings in relation to their environment (e.g., Levins and Lewontin 1985; Odling-Smee, Laland, and Feldman 2003; Sultan 2015).

We[2] propose that trees should be at the center of philosophical, environmental, and ethical debates, not only because they are a crucial—and undertheorized—life-form but also because trees pose significant philosophical and environmental challenges that must be addressed. While, as we will show, other life-forms pose similar challenges, trees expose the limits of our concepts and frameworks in the most direct and concrete way, thereby placing an explicit and urgent demand on us to think differently. By focusing on trees, our aim is to meet these challenges head-on and offer some insights into what we can learn from them.

To achieve this aim, we will consider not only the "why" question (*Why* trees?), but also the "how" question—*how* we represent, approach, and know trees, and whether our methods are appropriate. We will also consider how trees represent *themselves* and what we can learn from their form of self-presentation. By allowing trees to address us, we seek to think *with* trees about crucial environmental and ethical concepts—and thereby arrive at new insights into ourselves, our worlds, and our environmental futures.

2. *Taking Account of Trees*

We began this article with the desire to connect to Rimu—to enter Rimu's world in some way. But what exactly would it mean to *know* Rimu? What approaches can and should we use in our attempt to become familiar with the world of Rimu?

Questions concerning knowledge—how and what we know—have been central to environmental justice. This is because knowing a natural phenomenon is a condition for representing it, whether by *representation* we mean a legal or political category (as in representing someone in a court of law or in a governing body) or an epistemological category, whereby we seek to determine what something is in order to (eventually) determine its moral status. In both cases, representation is seen as necessary because it is by "speaking on behalf" of a natural phenomenon, or discerning its nature and capacities, that we can grasp and convey its needs and interests and, on that basis, argue for its legal, political, or moral status. Or, as Christopher D. Stone (1972: 464) put it in his influential (but problematic) article, representation is necessary to achieve environmental goals because the nonhuman other is "vegetal," "incompetent," and thus incapable of self-presentation.

Within the environmental humanities, representation often appears in the form of storytelling, as the narration of the life or way of life of nonhuman beings. This emphasis on narration is strategic because, as William Cronon (1992: 1374) put it, "Good stories make us care." By telling good stories—by relating the lives of nonhuman beings through narrative—Cronon claims, we may be able to transform habits of feeling and ways of thinking. This, in turn, can result in our becoming more (ethically, politically) engaged in the lives of more-than-human beings. Or, as Thom van Dooren (2014: 83) writes, by telling stories we begin to pay attention, and this allows us "to see differently, and so to be drawn into new kinds of relationships, new ethical obligations."

Of course, representing an other can be dangerous and can lead to problematic ethical positions. It often assumes—as Stone's language makes explicit—that the other is "incompetent," incapable of speaking on their own behalf. And this, as Eva Giraud (2019: 26) notes, can easily result in "forms of political ventriloquism that reinscribe inequalities rather than overturning them." Drawing on Donna Haraway, Giraud explains that the problem lies in the act of representation itself. As Haraway (1992: 313) argues, representing another "depends on possession of a passive resource . . . the silent object." It depends, in other words, on the view that the other is lacking in some way, a lack that is connected to an inability to speak on their own behalf.

In turn, storytelling might be—despite the best of intentions—a form of human imperialism. As Cronon himself notes, narrative is "a peculiarly human way of organizing reality." For this reason, he continues, environmental historians must be especially cautious in narrating the lives of the more-than-human world. Often these lives—and natural events more generally—do not follow the narratorial trajectory: they lack a "compelling drama" or a "protagonist," they are at times cyclical, and at times random (Cronon 1992: 1368). Furthermore, the ethical value of stories often rests on the ideal of empathic engagement. Such engagement, however, appears to be limited to those beings that are "like us": beings whose face and voice resembles our own, and whose needs and interests can be interpreted along the nexus of pain and pleasure (Marder 2012).[3]

According to Haraway, there is an alternative to what she calls the "political semiotics of representation"—namely, "articulation." In contrast to representation, which posits a knowing subject over against a known object, articulation is founded on "situated knowledges" and "points of view" (Haraway 1992: 309). Thus, rather than speaking *on behalf of another*, articulation speaks *with the other*. To speak *with* another assumes a different model of knowledge: one that, as Barad (2007: 379) puts it, is founded on "a direct material engagement, a practice of intra-acting with the world as part of the world in its dynamic material configuring, its ongoing articulation." In other words, knowledge need not be—indeed ought not to be—conceived along the lines of a subject-object relation. Rather, knowledge should be understood as contextualized, embodied, "entangled," practices of material engagement.

This has implications for how we tell stories. For if narrative is understood along similar lines, it follows that narrating nonhuman lives need not abide by the Aristotelian conception of narrative, with a clear beginning, middle, and end. After all, if we are "always already" participating in ongoing processes, then we are always "beginning in the middle" (as Friedrich von Schlegel [1958–2006, 2:178] had it). To tell stories, then, entails not simply describing a shared world but also describing dynamic processes whose beginnings and ends—and indeed whose very temporalities—do not map onto our human processes and temporalities.[4] Put differently, representation must aim to grasp and convey how the more-than-human other "organizes reality" *in their own way*: how living beings *construct* meaning and *experience* time—and thereby also construct and experience their world—in a way that does not align with human constructions of reality.

It is here, rather than in empathic engagement, that the ethical potential of storytelling lies. By attending to the lives of other beings, we become aware of their capacity to construct meaning—to participate in a meaningful, expressive world of their own—*and* of the fact that we are impinging on their world, often in destructive ways (Kohn 2013; van Dooren 2014). Accordingly, if the goal of representation is to lead us to think differently about our ethical relations and obligations, it follows that representation ought not simply to engage our empathy but also, and more importantly, it ought to discern and portray the ways in which more-than-human beings meaningfully engage with their meaning-laden worlds.

The question arises, however, as to whether telling such stories is possible in the case of trees. Is it possible, in other words, to "represent" Rimu, and trees more generally, without ascribing to them the role of incompetent or passive object, on the one hand, and, on the other, without infusing them with a human likeness that they do not possess (Plumwood 2009)? Is it possible to discern a "meaningful world-making" among trees? And what are the implications of these "tree stories"? Do they have the potential to influence our ways of thinking and acting—challenging us to reconsider our normative ideals?

Our aim, as an environmental philosopher and a plant physiologist, is to begin to answer these questions. To do so, we hope to work with, but also to further develop, the methodologies and aims of thinkers such as Haraway, Barad, Eduardo Kohn, and van Dooren, who have sought to tell stories of shared worlds, and—on that basis—to develop ethical positions. While these thinkers have not focused on trees or plants, we believe that by turning our attention to trees, we will build on and extend their methodology in important ways. For, as will become evident, tree stories cannot be told in the same way that animal stories are told. This is not only because—as noted above—trees do not have an evident "voice" or "face" with which we can identify or that we can easily interpret. Nor is it simply because trees are not as clearly "responsive" as animals, making our "response-ability" toward them—to follow Haraway's (2008) reasoning—less evident (see also Smith 2011; Krzywoszynska 2019; O'Brien 2020). It is also—and above all—because trees evidence a distinctive (and undertheorized) relationship to their environment, which calls into question any hard-and-fast distinction between living being and its world. As such, tree stories challenge narratives that rely on discerning how a living being relates to its world, or "constructs" a "meaningful world" for itself. For, as we will demonstrate, trees are *of* the world in the most immediate and fullest way, such that it is extremely difficult to say where an "individual" tree ends and where its "world" begins. This deep reciprocity is expressed in the tree's structure and development—giving us an important clue into the possibility of telling tree stories. For, despite apparent disadvantages (e.g., lacking a voice or face), the fact that trees are fundamentally expressive beings, whose stories (life stories, but also the stories of their world) are literally inscribed on their very form, means that trees are, in a significant sense, telling their own stories.

Accordingly, telling tree stories cannot amount to tracing the lived experience of a particular being or species in relation to its meaning-laden context but must involve exploring the expressive character of trees, the ways in which trees themselves tell their own stories. In turn, the implications of tree stories must also be different: by demanding us to rethink some of our most basic ideas about living beings and their worlds, tree stories challenge us to reconsider our understanding of ourselves, our environments, and our ethical ideals.

Our approach to telling tree stories and discerning their implications will draw on our two disciplines, relying on empirical research in order to develop a "picture" of the lives of trees. This picture is grounded in concrete (phenomenological) descriptions but aims to have *conceptual* significance (i.e., to amount to more than the accumulation of descriptions).[5] Put differently, this picture aims to generate insights into the character of trees and determine what this character tells us about living beings and environments more generally. Our aim is thus at once descriptive—to convey the lives of trees through "thick" descriptions grounded in empirical research—and conceptual, drawing attention to the ways in which trees call into question some of our most basic biological and ethical concepts and frameworks. In so doing, we hope to expand our understanding of storytelling in the environmental humanities *and* demonstrate how trees stories can lead to new conceptual insights and ethical frameworks. Let us, then, begin drawing this picture of trees using examples from a range of tree species and treed landscapes. We will also return to our friend Rimu to demonstrate crucial insights as they emerge.

3. The Phenomenon of Trees

Trees, like all plants, are rooted in the soil at a single location. They are essentially sessile, which means that their movements are usually at a much slower rate than the movement of animals. There are exceptional species of course: mimosa contracts its leaflets within seconds of being touched. But the big dynamic responses of trees are on time scales of hours (e.g., black locust leaves changing leaf angle over a day in response to the sun's angle [Liu et al. 2007]) or seasons (the fall of deciduous leaves in autumn).

Roots are the site of a second important feature of trees: many form symbiotic relationships with mycorrhizal fungi and nitrogen-fixing bacteria. These intra-kingdom relationships enhance the uptake of nutrients and water from the soil and allow atmospheric nitrogen fixation. The fungal partners also influence the rate of decomposition of organic matter, an important aspect of nutrient cycling in forests (Steidinger et al. 2019). For instance, ectomycorrhizal fungi, which dominate in warm and wet conditions, enhance decomposition, while arbuscular mycorrhizal fungi, which dominate in cool and dry conditions, inhibit decomposition rates. Hence, the tree's symbiotic partners regulate the rate at which nutrients are released into the soil and become available for growth.

Our friend Rimu forms an unusual relationship with an arbuscular mycorrhizal fungi, creating specialized nodules on fine roots to support the fungal partner. Arbuscular mycorrhiza are fungi that form structures for chemical exchange within the cells of plant roots. The large, spreading hyphal network, which is present with most arbuscular mycorrhiza symbioses (thought to increase access to soil nutrients such as phosphorus and nitrogen) is absent with mycorrhizal nodules in Rimu, so the benefit to Rimu of the partnership is not yet clear (Russell, Bidartondo, and Butterfield 2002). Most rimu trees grow on well-drained, fertile soils, but our friend grows on a water-logged peat soil with low nutrient availability. Any help in gaining scarce nutrients would be well worth Rimu's investment to build small nodule homes for its fungal partner.

Longevity is another distinguishing feature of trees. Trees are the longest life-form we know. "Methuselah," a bristlecone pine in eastern California, is at least 4,850 years old—it was already mature when Jesus Christ was born. "Sarv-e Abarkuh," the cypress of Abarkuh in Iran, is estimated to be a similar age to the oldest bristlecone pines and has been connected through legend to the ancient Iranian prophet Zoroaster. In Australia, a Huon Pine tree on Mount Read in Tasmania has been reproducing itself vegetatively at the same location for up to 10,000 years, and its live stems are easily more than 1,000 years old (Lloyd 2011). The longevity of trees leads to another feature that distinguishes them from birds and mammals: indeterminate growth. A tree may reach a maximum height, determined by a combination of genetic and environmental constraints, but will continue to increase in diameter until physical damage or disease eventually kill it (unless it reproduces vegetatively like the Huon Pine mentioned above, or "Pando," the quaking Aspen in Utah [DeWoody et al. 2008]).

Trees record their growth within their very structure, through the phenomenon of tree rings. Dendrochronology, or the study of tree rings as a record of age, emerged as a scientific technique through the work of A. E. Douglass (1919) about a century ago, when it was proven that the width of an annual ring is a record of growth over one year (Torbenson 2015). Wood formed during the spring growth-flush has large, thin-walled cells that are paler in color than the smaller, thicker-walled cells produced in late summer, resulting in a repeating pattern of concentric rings.[6]

As mentioned above, tree cores revealed that our friend Rimu is about four hundred years old. This means the tree has witnessed the rapid increase in human density and activity over its life so far: from the occasional Māori bird hunter or *pounamu* collector (*pounamu* are highly valued nephrite jade and bowenite stones, used to create tools, weapons, and ceremonial objects by Māori, and are found in waterways near Rimu) to the influx of Europeans during the West Coast gold rush in the 1860s, to the present-day dairy farmers.

Tree diameter measurements have shown us that trees grow fastest during years when they have ample water and sunlight and when temperatures are warmest (at least in the northern hemisphere temperate region where most of the original work was done), so tree-ring width has also been widely used to reconstruct past climates (e.g., Fritts 1974; Jacoby 2000).

However, tree rings record more than just the growth rate of an individual tree. By combining exact scientific measurements with an imaginative impulse, we gain insight not only into the history of a particular tree but also into the history of the planet as a whole.[7] The chemical composition of wood contains a chronological archive of the environment and the tree's response to that environment. The increasing atmospheric carbon dioxide concentration over the last one hundred years is recorded in the stable carbon isotope composition of tree rings because the carbon dioxide produced during fossil-fuel burning has fewer of the naturally occurring, but rare, carbon atoms with thirteen neutrons (Stuvier 1978; Francey and Farquhar 1982). This means that trees have an embodied record of both the Industrial Revolution and our current stubborn dependency on fossil fuels. Trees, in other words, might be best able to tell us exactly when anthropogenic climate change began to occur and be able to determine the most viable starting point of our current geological era, the Anthropocene.

In an inspired study, Chris Turney and coauthors took a core from the stem of a Sitka spruce (*Picea sitchensis*) planted on Campbell Island—a small remote island in the Southern Ocean—and found a sharp spike in the radiocarbon composition within the annual growth ring for 1965 (Turney et al. 2018). The peak reflects the fixation of atmospheric radiocarbon released during nuclear testing in the 1950s and 1960s. This, the scientists suggest, marks the beginning of the Anthropocene because it is the first time that an undeniably human-generated change in our earth's atmosphere has ever been recorded by the biosphere.

The tree's own response to environmental stresses is overlaid on the human transformation of biogeochemical cycles inscribed in its morphology. We can discern, for instance, that a tree went through a "stressful" growth season that resulted in a narrow annual ring, and we can interpret this as being caused by drought, or herbivory, or some other environmental or physiological effect. In fact, we can tell the tree's weekly story if annual rings are divided sequentially into thin slices and analyzed separately. Hot and dry summers are recorded as narrow growth rings with sharp peaks in carbon-13 and oxygen-18 composition, while mild, sunny summers with high rainfall result in broad-flat peaks and wide growth rings (Barbour, Walcroft, and Farquhar 2002).

Of course, all living beings record time. As biologist J. S. Haldane (1917: 98) put it in his book *Organism and Environment*, "In a living organism, the past lives on in the present" such that "it is literally true of life, and no mere metaphor, that . . . each moment of the past [is] in each moment of the present." In mammals, for instance, bones and tooth enamel are known to record a range of environmental and physiological signals during the time they formed. The radiocarbon concentration of femoral cortical bone (dense regions of thigh bones) has been used to determine the years during which terrestrial mammals such as brown bear and sika deer were in their adolescence (Matsubayashi and Tayasu 2019), and the Irish potato famine is recorded in the stable carbon and nitrogen isotope composition of human teeth and bones (Beaumont and Montgomery 2016).

However, these animal histories are generally integrative records formed during key developmental time periods, not chronological records like tree rings. Furthermore, the information they convey has more to do with the animal's particular history (its life cycle, its health and nutrition) and less to do with the world it inhabits. This, of course, is related to the fact that animals do not often remain in one place, such that the most widely used chronological histories in animals—fish inner ear bones—tell us less about the fish's environment and more about its life and well-being. (These tiny bones, called otoliths, grow at different rates over a year, resulting in annual rings just like tree rings, and their chemical composition can reveal details of fish migratory patterns and diet [Swanson 2017; Martino et al. 2019]).

In addition, while all living beings carry their past with them into their present and future selves, there is something distinctive and important about the ways in which trees do this. This has to do with the fact that trees embody their temporal history in a way that is far more explicit: it is literally inscribed in every one of their parts—from tree rings to growth pattern, to the tree's overall structure. Trees of the same species and age can look significantly different depending on their growth environment. When densely planted, trees grow long, straight trunks and small canopies, but when planted in a grass field, they grow shorter stems and broad crowns. The crown of a solitary oak spreads out in all directions, eventually achieving a dome shape, while an oak growing in a forest develops a small crown, and its growth is patterned on the growth of surrounding trees. Even within an individual tree, the leaves at the shady bottom of the canopy are anatomically different (larger and thinner) from those at the top (smaller and thicker). What this shows is that the structure and shape of a tree is a record or an expression of its environment and relationships. Put simply, trees express their context in their physical form.

Animals can, of course, reflect their environment. However, they do so in a far more limited way than plants and with far less plasticity. Trees reflect their context *in every one of their parts* (Holdrege 2013), exhibiting an openness toward the environment that we do not witness in animals. In humus-poor earth, it is not only the crown of the oak but also its root that expresses the environment: its root is short, with far less branching than the same species in humus-rich soil. Every one of its parts is ultimately telling the story of its distinctive context.

In a significant sense, trees are the living beings that manifest Haldane's insight on the historicity of life most clearly and concretely, placing a clear demand on us to think of life not as static and machinelike but as dynamic, context-sensitive, and plastic.

This, however, is not the extent of the implications of Haldane's statement. As recent work in theoretical biology has argued, Haldane is pointing to the fact that living beings are not fixed "substances" but dynamic "processes" (Nicholson and Dupré 2018). On this view, individual beings are expressions of multiple ongoing processes at varying scales. Trees—and indeed all plants—explicitly demonstrate this insight. This is because, as botanist Matthias Schleiden (1842–43, 1:31) put it, "in every moment of its life, the plant is only a part of itself." Trees are *always* developing, and this ongoing development is an essential aspect of their life cycle. When one part dies, another emerges: the flower withers and, in so doing, makes way for the fruit. The parts *anticipate* one another, and their developmental relations demand a distinctive perspective. We can only grasp the plant "whole" if we develop a certain form of attentiveness—one that is not focused on the present and that does not regard the tree as a static substance but is able to see how the various moments of development are moments of one process. Accordingly, plants, and trees especially, clearly manifest the developmental and historical character of life: they literally *embody* their history in every one of their constituent parts *and* their development character is expressed in their growth and in the relations between their parts.

Returning to Rimu, the broad diameter and soaring thirty-meter height of the tree speak of its age, robust health, and a favorable environment. But the physically recorded story of Rimu is not all about flourishing. The broken stump of a very large branch twenty meters above the ground records a violent and destructive storm in the recent past, despite the blurring cloak of epiphytic mosses.

Rimu and Rimu's neighbors have built-in flexibility to wind (in fact it is alarming to experience how much they move with the wind while standing on the rigid structure of the scaffolding access tower), so the storm that ripped the forty-centimeter-diameter branch from Rimu must have been extreme.

Extreme environmental conditions are recorded in Rimu, but Rimu and trees more generally also alter their environment. That is, trees are not merely receptive or passive in relation to their environment; they also transform their environments. Although a lot of work has gone into describing the ways in which animals construct their niches (Levins and Lewontin 1985; Odling-Smee, Laland, and Feldman 2003; Sultan 2015), less work has been done on how trees do so. Yet trees construct their environments at the largest scale, such that it is no exaggeration to say that trees do not simply transform their environments but *actively create* them. This is evident with respect to both of the "worlds" that trees inhabit: the soil below and the atmosphere above.

In forests, some tree species alter their environments in such radical ways that they determine the species composition around them. The giant kauri (*Agathis australis*), a species endemic to the northern regions of New Zealand, is one of the most sophisticated environmental engineers. Its fallen leaves create thick layers of humus on the forest floor (Jongkind, Velthorst, and Burman 2007). Over time, the highly acidic leachate from the humus can wash virtually all nutrients from the soil, resulting in a pale lens of low-nutrient, acidic soil within the root zone called a cup podzol. The plant communities growing on these highly modified soils are distinctly different to neighboring communities (Wyse, Burns, and Wright 2014). Trees, in other words, are fundamentally determining the structure and chemistry of the soil and—as such—are actively making their soil environment. Or, as F. John Odling-Smee, Kevin Laland, and Marcus Feldman (2003: 217) put it, where tree roots bind and stabilize bedrock, it is the trees that "literally hold the mountainside together."

The case is the same when we consider the atmosphere. In the vast forests of the Amazon, trees drive the hydrological cycle by lifting soil water into their canopies where it evaporates and is released to the atmosphere as vapor, a process called transpiration. Hence, much of the water that falls as rain in the Amazon comes from transpiration (estimated to be 30 to 50 percent), perhaps cycling a number of times from soil to atmosphere through trees before leaving the continent mostly via the massive river system (Salati et al. 1979). Furthermore, recent research in the southern Amazon has revealed that transpiration during the late dry season brings forward the dry-to-wet transition by two to three months (Wright et al. 2017). The dry season has been increasingly delayed in the southern Amazon in recent decades (Fu et al. 2013), prompting suggestions that continued clearing of land for agriculture and changes in fire regimes might trigger a collapse of the rainforest and the development of savannah (Staver, Archibald, and Levin 2011).[8]

This demonstrates that trees do more than simply influence or transform an already extant environment: they create it. There is thus no straightforward way by which to conceptualize the tree environment *before* or *without* the trees. The trees are an *expression* of their environment, as much as their environment is an *expression* of the trees, such that separating them—or seeking to understand them in contrast to one another—is to misunderstand their relation. The Amazon is an expression of the trees that make it up and regulate its hydrological cycles, just as the trees are an expression of this specific context. Without trees, the climate and atmosphere of the Amazon would fundamentally alter, and without these cycles, the trees would cease to exist.

What trees reveal—more so than any other living being—is that the organism-environment relationship is one of deep reciprocal causation and dependency. While all living beings express and influence their environments, trees are shaped by their environments to a greater extent, exhibiting a degree of openness and plasticity that is not evident in animal bodies. Furthermore, while all living beings play a role in transforming their environments, trees do so more subtly (in ways that are less visible but also crucial—think of soil) and at a greater scale (think of the Amazon), such that we cannot think of an "environment" without (or before) its trees (or lack thereof). Ultimately, while we can conceive of an animal as standing in contrast to (even in opposition to) its environment, this is impossible in the case of trees.[9] This is because the "environment" is realized in and through the activities of the trees, while trees—in their morphology, in adjustments they make to tissue allocation and root deployment, and in the ways successive generations are affected by factors in parent environments ("cross-generational plasticity" [Sultan 2003])—are expressions of this environment. The one does not precede and effect the other (the environment does not preexist the tree, nor vice versa). They emerge in relation to one another. In trees, then, we find the most convincing and thorough example of what biologist Sonia Sultan (2015: 31) has described as the "causally multidirectional interactions" between organism and environment: interactions that fundamentally blur the boundary between the two.

Trees thus concretely demonstrate the historicity and dynamism of life (Haldane 1917) *and* the intimate and nonlinear relation between living beings and their worlds (Sultan 2015). They reveal that living beings are not static "substances" but ongoing processes participating in and contributing to other processes. As such, they do not stand over and against their environments, as if these environments are static backdrops of plant and animal activity. Rather, they collaborate with and are inseparable from their worlds. Consider the Huon Pine on Mount Read or the aspen forests mentioned above, which are composed of clones connected underground through far-reaching roots (Clarke 2010). In both cases, it is extremely difficult to determine where the plant "substance" ends and where its "world" begins—whether in terms of its life cycle or in its relation to its (environmental) context.

We suggest that the concept of "embodied history" captures these various characteristics—illuminating the developmental process character of trees and their intimate and reciprocal relationship to their environments. By "embodied history" we mean the ways in which trees are *embodiments* or *expressions* of their environments and the ways in which environments are also *expressions* of their trees. What the notion of embodied history aims to convey is that trees and environments are inextricably connected—a connection that is literally inscribed on their bodies, making explicit the absolute dependency of the one on the other. Furthermore, embodied history seeks to call attention to the fact that the relation between tree and environment is an *expressive* relation, one in which trees are expressions of their environments and are also expressed in and through their environments.

This way of recasting tree relationships and identities has important consequences for our understanding of representation. For what the embodied history of trees shows is that trees are, in a significant sense, expressive beings, and that they express *both* themselves *and* their world. Trees, in other words, *tell their stories*. If we pay attention to trees, to their ways of being and becoming, we can "read" their stories in every one of their parts—from the bark to the roots and crown. But we can do more: we can also read the stories of their worlds. For tree stories are nondualistic in that trees tell not only their (individual) stories but also—and always—the stories of their worlds, the worlds from which they cannot be clearly (or substantially) separated. The embodied history of trees can, in turn, teach us to tell stories of a shared and dynamic world, a world in which living beings express their "environments" just as much as their "environments" express them. These stories reveal a mutually transformative relation, where living being and world form and inform one another in the deepest and most encompassing of ways, and in which development occurs bilaterally. To tell the story of one tree is also to tell the story of the various beings and relations that enable and support this tree—and which this tree enables and supports. It is to tell a story of an ongoing dynamic process of relation over decades and centuries.

This alone challenges some of our most basic assumptions about living beings and their worlds—assumptions that sit at the heart of environmental ethics. Accordingly, we are led to wonder how the embodied history of trees (as a concept) and tree stories (as expressive of a mutual developmental process) can help us to develop a more encompassing form of environmental ethics—one that does not revolve around discrete individuals and environments, and one that is thereby not hindered by either-or scenarios and ways of thinking. Put differently, can the embodied history of trees help us to move beyond some of the aporias that continue to challenge environmental ethics? Let us turn to these questions.

4. The Embodied History of Trees and More-than-Human Ethics

Until recently, trees (and indeed plants more generally) have received little attention within environmental ethics (Kallhoff 2014; Kallhoff, Di Paola, and Schörgenhumer 2018). This may be because environmental ethics has largely focused on beings that are "like us" (Regan 1983)—whether it is because they are "sentient" and suffer (Singer 1975) or because they possess certain capabilities (e.g., for social life or for feeling joy and sadness) (Nussbaum 2006), or because they can "evaluate" their world as good or bad for them (Korsgaard 2018).

On these views, moral consideration is understood as something that we must *extend* from the human to the more-than-human. This extension occurs in two ways: (a) the extension of human *values* (e.g., "rights") to the more-than-human world; and (b) the extension of the human *form* to the more-than-human world. This, in turn, leads to an understanding of what a moral subject is: an entity whose form somehow mirrors or reflects the human form and is thus recognizably separable from its world and others. Or, as Martha Nussbaum (2000: 78) put it, a moral agent is "able to move freely from place to place; [and whose] bodily boundaries [are] treated as sovereign."

There are, of course, ethical

perspectives that seek to be more capacious. Biocentric views, for instance, argue that all biological beings—whether they resemble the human form or not—have "interests" and are thus deserving of moral consideration (Agar 2001). Proponents of ecological holism and ecocentricism, in turn, often contend that ecosystems express interests independent of the interests of the living beings that make them up (Callicott 1989; Newman, Varner, and Linquist 2017: 295). Or they argue that any being that can be designated as "individual" (where the term applies both to individual entities and to ecosystems) deserves moral consideration (Varner 1985). However, even these perspectives often fall short when it comes to trees. This is because trees cannot be separated from their environments in any substantial way, such that any distinction between organism and environment necessarily excludes trees. One cannot, for instance, wish to "save" a tree but not the forest, or the tree but not the soil. In thinking about trees, it is imperative to consider *not only* the individual tree or species, or even the many species that make up a particular context (i.e., the usual manner by which trees, and living beings more generally, are discussed), but *also* to consider the various processes—organic and inorganic, small-scale and large-scale, present and futural—that support the tree(s) and that the tree(s) itself (themselves) support(s).

Indeed, by taking the embodied history of trees seriously and seeking to be guided by it in our moral considerations, a different set of concepts—and a different framework—comes to the fore. In the place of an extensionist ethics that grounds moral consideration in human values (e.g., rights) or the human form, we begin to develop an ethics founded on what trees tell us about their lives and ways of being. In the place of autonomy and sovereignty, separateness and difference, we find development and process, sensitivity and openness, dialogue, plasticity, longevity, and rootedness. While each of these features is present in all living beings, they are most vividly manifest in trees. Thus, in calling attention to them, we do not imply that trees are exceptional but that they are exemplary: they exemplify what is at the heart of all living beings and environments and thereby offer unparalleled lessons for understanding what it means to be alive and what it takes to understand living beings and environments. We will discuss each of these concepts in turn.

Living beings are developmental processes. The very possibility of being alive depends on a number of physiological processes (from digestion to growth), and individual living beings go through a developmental life cycle. In turn, generations develop through the process of heredity, and species undergo evolutionary development. While these aspects are true of all living beings, they are most vividly embodied in trees, which are *always in development,* insofar as they are *never fully* completed—there is always one part that is not yet developed or that is absent (e.g., the flower or the fruit). Because they are never "whole" at any one time, we cannot conceive of trees as static, discrete substances. In other words, we cannot commit the mistake that is easy to commit in the case of animals. For what trees clearly demonstrate is that living beings are not finished *products* but *productivities*—sites of self-production.

Tree development involves the formation of collaborations, unions, and alliances across (species) boundaries. This includes symbiotic relationships with other species as well as the fundamental relation between tree and environment. This relationship—which goes hand in hand with the developmental character of trees—is only possible because trees are sensitive and open to their environments (they either remain open, sensitive, and responsive, or they die). Their development is dependent on the ongoing dialogues that occur between them and their dialogue partner(s)—whatever or whomever they may be.

The openness that trees exhibit does not imply a passive form of receptivity but an active one. Trees effect change in their world—change that is neither haphazard nor one-sided. It is not haphazard in that the changes which trees effect are outcomes of an ongoing dialogue. Trees are not *reacting* in a random or sudden way to their environments. Rather, their responses are studied and emerge out of a developed sense of their environments. One can thus say that trees respond to their environments in an intentional way—where *intentional* does not mean (as it does in the human context) action done in line with prior rational deliberation, but means *nonarbitrary*. There is nothing arbitrary about their response, in that it comes through years of continued relating and sensing. The change is, furthermore, not one-sided, in that the tree does not force its intentions on the environment but achieves its goals through collaboration with its environment. Trees develop dialogically, through a continuing (chemical) conversation with their surroundings, which include other plant, microbe, and animal species.

In their ability to collaborate with their environments, trees exhibit an incredible plasticity. From one perspective, this might be regarded negatively. However, this plasticity goes hand in hand with their openness, sensitivity, and dialogical character. Trees do not stand as completed objects over against an opposed world but as living dynamic beings acting with and in relation to their living evolving world.

Attending to trees in this way generates an ethical principle by which living beings can be considered and their well-being determined. We must assume a *developmental perspective*, which does not only concern how they are now but also considers how they have been and how they will be in the future; a perspective that does not only focus on their present relations but also recognizes these relations as sites of development and transformation.

The long lives of many tree species confirm the developmental perspective, adding an important layer to what it entails. An anthropocentric environmental ethics framework or an ethics that focuses only on mammals and birds can very easily overlook questions concerning life span and simply focus on the present or near-past and near-future. An ethical perspective founded on the lives of trees, by contrast, must involve consideration of the life span and dynamic, relational phenomenon of trees, or what we might describe as "long thinking."

Trees are rooted, and while they move, their movement is minimal beyond growth. This means, first, that trees can hardly be separated from the world that they inhabit: where the individual tree begins and where its environment ends is almost impossible to delimit. It also means that they cannot be understood as autonomous agents whose sovereignty aligns with their (physical) separation from other beings. In the place of autonomy, sovereignty, and separateness, trees present us with embeddedness and collaboration.

The ethical concepts and framework that emerge here approximate and complement the positions of deep ecology, relational, feminist, and Indigenous environmental ethics (Naess 1989; Barad 2007; Haraway 2008; Graham 2014; Brigg, Graham, and Weber 2021; Tynan 2021; see also the statement made by Indigenous Peoples of Mother Earth [2012] assembled at the site of Kari-Oca). While these various perspectives differ, they share the view that relationality—rather than autonomy or rationality—is the ground for moral responsibility. Thus, in contrast to the claim that responsibility involves two autonomous agents (Strawson 1962) who are reciprocally responsible (Watson 2014), thinkers from these traditions have argued that responsibility *emerges* in and through relations. Or, as Shawn Wilson (2008: 7) puts it, responsibility is "being accountable to your relations."

We suggest that the embodied history of trees concretely exemplifies the insights developed in these frameworks, providing some of the most striking examples of nonautonomous agency, reciprocal relationality, sensitivity, responsiveness, and entanglement. But the embodied history of trees also adds to these insights in important ways. Specifically, the embodied history of trees places emphasis on development, process, and life cycle, expanding our moral considerations beyond the present or the short-term and demanding that we think about the moral status of *developmental processes* rather than individuals or environments. Let us explain.

Trees, as we have sought to demonstrate, are developmental processes that are inextricably part of their larger, and multiple, contexts. Furthermore, they belong to two "worlds" at once: the soil beneath and the atmosphere above. Both of these features highlight the limits of thinking about *either* living beings *or* environments in an exclusionary way. As developmental processes, trees are always of multiple worlds (or, better: processes), from which they cannot be separated; they are both members of a species and collaborators within larger processes; they are both present and past as well as futural. To conceive of them in separation from their worlds or the processes of which they are part, to seek to determine their interests at any point in time irrespective of their ongoing development, is to misunderstand them. What the embodied history of trees shows, then, is that we must think not only in terms of relations and entanglements but also in terms of ongoing processes and dynamic histories. In other words, we must think of relations not as fixed entities but as developmental and processual—as involving many other ongoing processes into the future. Accordingly, we must think dynamically not only about living beings and environments but also about their relations—the relations of which they are part and that they themselves are.

This, we submit, is a crucial lesson that the embodied history of trees imparts, and it should be front and center of environmental thinking. To develop an environmental ethics grounded in tree stories is to develop an environmental ethics that takes seriously the dynamic, developmental character of living beings and environments, of relations and entanglements, and regards all life as an ongoing mutually supportive process deserving of long-term moral consideration.

Our time standing with Rimu has taught us that we cannot tell the story of an individual tree as a linear narrative with a main character striving toward a particular outcome at a point in time. Rather, Rimu must be represented through "long thinking": as inextricably connected, exquisitely sensitive, developmental, dynamic, and in constant dialogue with its biotic and abiotic environment. If we humans were able to construct social and political systems based on such an environmental ethic (i.e., a truly multispecies consideration of environmental ethics and justice), we would not be struggling to solve the wicked problems of climate change, pollution, and environmental degradation. But perhaps a more modest first step would be to seek to *emulate* the embodied history of trees in our own thinking and acting—to *retrain* our perception and judgment to become more sensitive and dynamic, more rooted and involved in our worlds, and more attuned to our ever-evolving relations and the processes of which we are part. In short: to think and act like a tree.

Acknowledgments

The authors gratefully acknowledge the comments from Danielle Celermajer, Sophie Chao, and the anonymous reviewers, which have significantly improved the manuscript. We also thank Te Kahautu Maxwell for his generosity in sharing mātauranga Māori regarding the *whakapapa* of trees and people.

Notes

1. Here we are using "Rimu" as a proper name for a specific individual tree of the species *Dacrydium cupressinum*, commonly known as rimu.
2. The "we" here—and throughout the article—refers to the two authors.
3. As Michael Marder notes, in the phenomenological tradition, beginning with Husserl, we witness an increasing interest in empathy as foundational for intersubjective experience. However, empathy has significant limitations when it comes to plants. While Husserl outright rejects the possibility of empathy with nonhuman beings, Edith Stein's important work on empathy allows for this possibility. This is because Stein (unlike Husserl) grounds empathy in the bodily subject (or living body, *Leib*) (Stein 2021: 264–67). We are thus able to experience empathic projection with other living bodies. However, as Marder elaborates, for Stein embodiment is understood on the basis of the human form, such that with decreasing similarity between the human form and the more-than-human form, we also witness decreasing possibility for empathic engagement. Embodiment as elaborated in this article departs from the conception of embodiment in this sense and is understood as describing the character of trees as embodied beings, who also embody their history. In this way, it is closer to other thinkers in the phenomenological tradition, from Johann Wolfgang von Goethe (see n. 5) to Maurice Merleau-Ponty (2003: 174), who speaks of "sensible ideas" to describe the goal of unifying thick phenomenological descriptions of phenomena with conceptual insight. For an account of Merleau-Ponty's notion of sensible ideas in relation to Goethe's phenomenology, see Fischer 2021.
4. When we use *our* here, we are aware that the implications are far too general and do not accurately depict the diversity of human cultures. The intention is to convey the Western conceptualization of narrative and temporal experience as developed by Aristotle and assumed by Cronon.
5. As indicated above (see n. 3), our approach is inspired by the protophenomenological approach practiced by Goethe (1749–1832) and also Alexander von Humboldt (1769–1859), who was influenced by Goethe. While Goethe developed the study of form (i.e., morphology) and sought to describe the "transforming form" of plants through the notion of metamorphosis, Humboldt developed a science that he called *physiognomy*, which seeks to trace the gestures, expressions, and actions of living beings in order to discern the "character" of a region. Furthermore, Humboldt articulated the idea of a *Naturgemälde*, which literally means "a portrait of nature," in order to depict—portray—the essential relations among living beings and environments. Both Goethe and Humboldt emphasize a continuity between thick description and explanation, and (in their empirical and theoretical work) show how we can arrive at conceptual insights in and through descriptions—i.e., how we can (and indeed must) remain *with* the phenomena

rather than depart from them. As Goethe puts it, "The phenomena themselves are the theory." In turn, their respective studies of morphology and physiognomy emphasize expressiveness, visibility, and encounter. For a discussion of Goethe's phenomenological approach to the study of nature, see Seamon 2005; Seamon and Zajonc 1998; Hennigfeld 2015. For a discussion of Humboldt's, see Nassar 2022.

6. The translation we humans impose on the voice of the tree recorded within the tree rings is the knowledge that in many tree species, the time period of cell formation from the start of one light-colored band of cells to the start of the next is one season (i.e., one year).
7. This is how Valerie Trouet (2020) describes it.
8. This point was made already in the early nineteenth century by Alexander von Humboldt following his observations of tree razing by European colonists at Lake Valencia in present-day Venezuela. The colonizers had specifically chosen an area with a good amount of annual rain for their agriculture; however, once the forests were razed, the rain also went—making for a very different (dry) environment, which was not conducive to agriculture. See Humboldt 1816–26: 72.
9. The very notion of an "organism" is generally understood to describe a being that stands in contrast to its environment—hence the designation of "organism *and* environment" (see, for instance, Mossio and Moreno 2010; Nicholson 2014). This is because, first, the organism must be capable of undertaking independent processes, which distinguish it from its environment (e.g., metabolism), which implies some form of closure, and "autonomy" from its environment. Second, it has to do with evolution, because evolution depends on the idea that biological individuals develop "fitness" through interactions with this external entity—that is, the environment. For the problems that this conceptualization of the organism faces when we turn to plants, see Clarke 2010.

References

Agar, Nicholas. 2001. *Life's Intrinsic Value*. New York: Columbia University Press.

Barad, Karen. 2007. *Meeting the Universe Halfway: Quantum Physics and the Entanglement of Matter and Meaning*. Durham, NC: Duke University Press.

Barbour, M. M., A. S. Walcroft, and G. D. Farquhar. 2002. "Seasonal Variation in $\delta^{13}C$ and $\delta^{18}O$ of Cellulose from Growth Rings of *Pinus radiata*." *Plant, Cell, and Environment* 25, no. 11: 1483–99.

Beaumont, Julia, and Janet Montgomery. 2016. "The Great Irish Famine: Identifying Starvation in the Tissues of Victims Using Stable Isotope Analysis of Bone and Incremental Dentine Collagen." *PLOS ONE* 11, no. 8: e0160065. https://doi.org/10.1371/journal.pone.0160065.

Brigg, Morgan, Mary Graham, and Martin Weber. 2021. "Relational Indigenous Systems." *Review of International Studies*, August 10. https://doi.org/10.1017/S0260210521000425.

Callicott, J. Baird. 1989. *In Defense of the Land Ethic*. Albany: State University of New York Press.

Clarke, Ellen. 2010. "Plant Individuality and Multilevel Selection Theory." In *The Major Transitions in Evolution Revisited*, edited by Brett Calcott and Kim Sterelny, 227–50. Cambridge, MA: MIT Press.

Coccia, Emanuele. 2019. *The Life of Plants*. London: Polity.

Cronon, William. 1992. "A Place for Stories: Nature, History, and Narrative." *Journal of American History* 78, no. 4: 1347–76.

DeWoody, Jennifer, Carol A. Rowe, Valerie D. Hipkins, and Karen E. Mock. 2008. "'Pando' Lives: Molecular Genetic Evidence of a Giant Aspen Clone in Central Utah." *Western North American Naturalist* 68, no. 4: 493–97.

Douglass, A. E. 1919. *Climatic Cycles and Tree Growth*. Vol. 1. Washington, DC: Carnegie Institute of Washington.

Fischer, Luke. 2021. "A Poetic Phenomenology of the Season." in *The Seasons: Philosophical, Literary, and Environmental Perspectives*, edited by Luke Fischer and David Macauley, 68–92. Albany: State University of New York Press.

Francey, R. J., and G. D. Farquhar. 1982. "An Explanation of 13C/12C Variations in Tree Rings." *Nature* 297: 28–31.

Fritts, Harold C. 1974. "Relationships of Ring Widths in Arid-Site Conifers to Variations in Monthly Temperature and Precipitation." *Ecological Monographs* 44, no. 4: 411–40.

Fu, Rong, et al. 2013. "Increased Dry-Season Length over Amazonia in Recent Decades and Its Implication for Future Climate Projection." *Proceedings of the National Academy of Sciences* 110, no. 45: 18110–15.

Giraud, Eva Haifa. 2019. *What Comes after Entanglement? Activism, Anthropocentrism, and an Ethics of Exclusion*. Durham, NC: Duke University Press.

Graham, Mary. 2014. "Aboriginal Notions of Relationality and Positionalism." *Global Discourse* 4, no. 1: 17–22.

Haldane, John S. 1917. *Organism and Environment as Illustrated by the Physiology of Breathing*. New Haven, CT: Yale University Press.

Hall, Matthew. 2011. *Plants as Persons*. Albany: State University of New York Press.

Haraway, Donna. 1992. "The Promises of Monsters: A Regenerative Politics for Inappropriate/d Others." In *Cultural Studies*, edited by Lawrence Grossberg, Cary Nelson, and Paula A. Treichler, 295–337. New York: Routledge.

Haraway, Donna. 2008. *When Species Meet*. Minneapolis: University of Minnesota Press.

Head, Lesley, Jennifer Atchison, Catherine Phillips, and Kathleen Buckingham. 2014. "Vegetal Politics: Belonging, Practices, and Places." *Social and Cultural Geography* 15, no. 8: 861–70.

Helmreich, Stefan. 2009. *Alien Ocean: Anthropological Voyages in Microbial Seas*. Berkeley: University of California Press.

Hennigfeld, Iris. 2015. "Goethe's Phenomenological Way of Thinking and the Urphänomen." *Goethe Yearbook* 22: 143–67.

Holdrege, Craig. 2013. *Thinking Like a Plant*. Great Barrington, MA: Lindisfarne.

Humboldt, Alexander von. 1816–26. *Voyage aux régions* équinoxiales *du noveau continent, fait in 1799, 1800, 1801, 1802, 1803, et 1804*. Quatro: 3 vols: vol. 1, Paris: F. Schoell; vol. 2, Paris: N. Maze; vol. 3, Paris: Gide Fils. Octavo: 13 vols.: vols. 1–4, Paris: Libraria Graeco-Latino-Allemande; vols. 5–8, Paris: N. Maze; vols. 9–13: Paris: Smith/Gide.

Indigenous Peoples of Mother Earth. 2012. "Kari-Oca 2 Declaration: Indigenous People's Global Conference on Rio+20 and Mother Earth." Indigenous Environmental Network, June 17. https://www.ienearth.org/kari-oca-2-declaration/.

Jacoby Gordon C. 2000. "Dendrochronology." In *Quaternary Geochronology: Methods and Applications AGU Reference Shelf*, vol. 4, edited by Jay Stratton Noller, Janet M. Sowers, and William R. Lettis, 11–20. Washington, DC: American Geophysical Union.

Jones, Owen, and Paul Cloke. 2002. *Tree Cultures*. Oxford: Berg.

Jongkind, A. G., E. Velthorst, and P. Burman. 2007. "Soil Chemical Properties under Kauri (*Agathis australis*) in the Waitakere Ranges, New Zealand." *Geoderma* 141, nos. 3–4: 320–31.

Kallhoff, Angela. 2014. "Plants in Ethics: Why Flourishing Deserves Moral Respect." *Environmental Values* 23, no. 6: 685–700.

Kallhoff Angela, Marcello Di Paola, and Maria Schörgenhumer, eds. 2018. *Plant Ethics: Concepts and Applications*. Abingdon, UK: Routledge.

Kohn, Eduardo. 2013. *How Forests Think: Toward an Anthropology Beyond the Human*. New York: Columbia University Press.

Korsgaard, Christine M. 2018. *Fellow Creatures: Our Obligations to the Other Animals*. Oxford: Oxford University Press.

Krzywoszynska, Anna. 2019. "Caring for Soil Life in the Anthropocene." *Transactions of the Institute of British Geographers* 44, no. 4: 661–75.

Levins, Richard and Richard Lewontin. 1985. *The Dialectical Biologist*. Cambridge, MA: Harvard University Press.

Liu, Cheng-Cheng, Clive V. J. Welham, Xian-Qiang Zhang, and Ren-Qing Wang. 2007. "Leaflet Movement of *Robinia pseudoacacia* in Response to a Changing Light Environment." *Journal of Integrative Plant Biology* 49, no. 4: 419–25.

Lloyd, Graham. 2011. "The Oldest Tree." *Australian*, September 10.

Marder, Michael. 2012. "The Life of Plants and the Limits of Empathy." *Dialogue* 51, no. 2: 259–73.

Marder, Michael. 2013. *Plant-Thinking: A Philosophy of Vegetal Life*. New York: Columbia University Press.

Martino, Jasmine C., Anthony J. Fowler, Zoë A. Doubleday, Gretchen L. Grammer, and Bronwyn M. Gillanders. 2019. "Using Otolith Chronologies to Understand Long-Term Trends and Extrinsic Drivers of Growth in Fisheries." *Ecosphere* 10, no. 1: e02553. https://doi.org/10.1002/ecs2.2553.

Matsubayashi, Jun, and Ichiro Tayasu. 2019. "Collagen Turnover and Isotopic Records in Cortical Bone." *Journal of Archaeological Science*, no. 106: 37–44.

Merleau-Ponty, Maurice. 2003. *Nature: Course Notes from the Collège de France*. Translated by Robert Vallier. Evanston, IL: Northwestern University Press.

Mossio, Matteo, and Alvaro Moreno. 2010. "Organisational Closure in Biological Organisms." *History and Philosophy of the Life Sciences* 32, nos. 2–3: 269–88.

Myers, Natasha. 2020. "How to Grow Liveable Worlds: Ten (Not-So-Easy) Steps for Life in the Planthroposcene." *ABC Religion and Ethics*, January 28. https://www.abc.net.au/religion/natasha-myers-how-to-grow-liveable-worlds:-ten-not-so-easy-step/11906548.

Naess, Arne. 1989. *Ecology, Community, Lifestyle*. Cambridge: Cambridge University Press.

Nassar, Dalia. 2022. *Romantic Empiricism: Nature, Art, and Ecology from Herder to Humboldt*. New York: Oxford University Press.

Newman, Jonathan A., Gary Varner, and Stefan Linquist 2017. *Defending Biodiversity: Environmental Science and Ethics*. Cambridge: Cambridge University Press.

Nicholson, Daniel J. 2014. "The Return of the Organism as a Fundamental Explanatory Concept in Biology." *Philosophy Compass* 9, no. 5: 347–59.

Nicholson, Daniel J., and John Dupré. 2018. *Everything Flows: Towards a Processual Philosophy of Biology*. Oxford: Oxford University Press.

Nussbaum, Martha C. 2000. *Women and Human Development: The Capabilities Approach*. Cambridge: Cambridge University Press.

Nussbaum, Martha C. 2006. *Frontiers of Justice: Disability, Nationality, Species Membership*. Cambridge, MA: Harvard University Press.

O'Brien, Anne Therese. 2020. "Ethical Acknowledgement of Soil Ecosystem Integrity amid Agricultural Production in Australia." *Environmental Humanities* 12, no. 1: 267–84.

Odling-Smee, F. John, Kevin N. Laland, and Marcus W. Feldman. 2003. *Niche Construction: The Neglected Process in Evolution*. Princeton, NJ: Princeton University Press.

Plumwood, Val. 2009. "Nature in the Active Voice." *Australian Humanities Review*, no. 46: 113–29.

Regan, Tom. 1983. *The Case for Animal Rights*. Berkeley: University of California Press.

Russell, Angela J., Martin I. Bidartondo, and Brian G. Butterfield. 2002. "The Root Nodules of the Podocarpaceae Harbour Arbuscular Mycorrhizal Fungi." *New Phytologist* 156, no. 2: 283–95.

Salati, Eneas, Attilio Dall'Olio, Eiichi Matsui, and Joel R. Gat. 1979. "Recycling of Water in the Amazon Basin: An Isotopic Study." *Water Resource Research* 15, no. 5: 1250–58.

Schlegel, Friedrich von. 1958–2006. *Kritische-Friedrich-Schlegel Ausgabe*. Edited by Ernst Behler, Jean Jacques Anstett, and Hans Eichner. 23 vols. Paderborn: Schöningh.

Schleiden, M. J. 1842–43. *Grundzüge der wissenschaftlichen Botanik*. 2 vols. Leipzig: Wilhelm Engelmann.

Seamon, David. 2005. "Goethe's Way of Science as a Phenomenology of Nature." *Janus Head* 8, no. 1: 86–101.

Seamon, David, and Arthur Zajonc, eds. 1998. *Goethe's Way of Science: A Phenomenology of Nature*. Albany: SUNY Press.

Singer, Peter. 1975. *Animal Liberation*. New York: Avon.

Smith, Mick. 2011. "Dis(appearance): Earth, Ethics, and Apparently (In)significant Others." *Australian Humanities Review*, no. 50: 23–44.

Staver, A. Carla, Sally Archibald, and Simon A. Levin. 2011. "The Global Extent and Determinants of Savanna and Forest as Alternative Biome States." *Science* 334, no. 6053: 230–32.

Steidinger, B. S., et al. 2019. "Climatic Controls of Decomposition Drive the Global Biogeography of Forest-Tree Symbioses." *Nature* 569: 404–8.

Stein, Edith. 2021. "Selections from *On Empathy*." In *Women Philosophers in the Long Nineteenth Century: The German Tradition*, edited by Dalia Nassar and Kristin Gjesdal, 248–72. New York: Oxford University Press.

Stone, Christopher D. 1972. "Should Trees Have Standing? Toward Legal Rights for Natural Objects." *Southern California Law Review* 45, no. 2: 450–501.

Strawson, Peter F. 1962. "Freedom and Resentment." *Proceedings of the British Academy*, no. 48: 187–211.

Stuvier, Minze. 1978. "Atmospheric Carbon Dioxide and Carbon Reservoir Changes." *Science* 199: 253–58.

Sultan, Sonia E. 2003. "Phenotypic Plasticity in Plants: A Case Study in Ecological Development." *Evolution and Development* 5, no. 1: 25–33.

Sultan, Sonia E. 2015. *Organism and Environment*. Oxford: Oxford University Press.

Swanson, Heather Anne. 2017. "Methods for Multispecies Anthropology: Thinking with Salmon Otoliths and Scales." *Social Analysis* 61, no. 2: 81—99.

Torbenson, Max C. A. 2015. "Dendrochronology." In *Geomorphological Techniques*, edited by L. E. Clark and J. M Neild, chap. 4, sec 2.8, p. 1. British Society for Geomorphology.

Trouet, Valerie. 2020. *Tree Story: The History of the World Written in Rings*. Baltimore: Johns Hopkins University Press.

Troy, Jakelin. 2019. "Trees Are at the Heart of Our Country—We Should Learn Their Indigenous Names." *Guardian*, April 1. https://www.theguardian.com/commentisfree/2019/apr/01/trees-are-at-the-heart-of-our-country-we-should-learn-their-indigenous-names.

Turney, Chris S. M., et al. 2018. "Global Peak in Atmospheric Radiocarbon Provides a Potential Definition for the Onset of the Anthropocene Epoch in 1965." *Scientific Reports* 8.3293.

Tynan, Lauren. 2021. "What Is Relationality?" *Cultural Geographics* 28, no. 4: 597–610.

van Dooren, Thom. 2014. *Flight Ways: Life and Loss at the Edge of Extinction*. New York: Columbia University Press.

Varner, Gary. 1985. "The Schopenhauerian Challenge in Environmental Ethics." *Environmental Ethics* 7, no. 3: 209–30.

Wallach, Arian D., Marc Bekoff, Chelsea Batavia, Michael Paul Nelson, and Daniel Ramp. 2018. "Summoning Compassion to Address the Challenges of Conservation." *Conservation Biology* 32, no. 6: 1255–65.

Wandersee, James H., and Elisabeth E. Schussler. 1999. "Preventing Plant Blindness." *American Biology Teacher* 61, no. 2: 82–86.

Watson, Gary. 2014. "Peter Strawson on Responsibility and Sociality." In *Oxford Studies in Agency and Responsibility*, vol. 2, edited by David Shoemaker and Neal Tognazzini, 15–32. Oxford: Oxford University Press.

Wilson, Shawn. 2008. *Research Is Ceremony: Indigenous Research Methods*. Black Point, NS: Fernwood.

Wright, Jonathan S., Rong Fu, John R. Worden, Sudip Chakraborty, Nicholas E. Clinton, Camille Risi, Ying Sun, and Lei. Yin. 2017. "Rainforest-Initiated Wet Season Onset." *Proceedings of the National Academy of Sciences* 114, no. 32: 8481–86.

Wyse, Sarah V., Bruce R. Burns, and Shane D. Wright. 2014. "Distinctive Vegetation Communities Are Associated with the Long-Lived Conifer *Agathis australis* (New Zealand kauri, Araucariaceae) in New Zealand Rainforests." *Austral Ecology* 39, no. 4: 388–400.

Dalia Nassar is associate professor of philosophy at the University of Sydney. Her work sits at the intersection of the history of philosophy and science, aesthetics, the philosophy of nature, and environmental philosophy. She is the author of, most recently, *Romantic Empiricism: Nature, Art, and Ecology from Herder to Humboldt* (2022).

Margaret Barbour is professor of plant physiology and dean of Te Aka Mātuatua—School of Science at the University of Waikato. She studies the interaction between plants and their environment from a Western science perspective, with a particular focus on stable isotopic techniques.

MEDITATIONS on WRITING HELL

Hayley Singer

Abstract This essay is a broken elemental thing composed of cuts, by which is meant outtakes. Outtakes are scenes or sequences that never make it into a film. The scenes collected here have been retrieved from the cutting floor of the editing suite in its author's mind and reassembled in ways that hold onto an ambitious claim—to think of narrative cuts and silences as interruptive forces in the operation of writing and the imaginative rendering of the abattoir. Working with outtakes helps the author approach, in a new way, questions the author has been exploring for a while now: How can writers critically respond to the existence of abattoirs? What strategies might writers engage to render normalized forms of violence against animals strange and even intolerable through particularly literary practices, strategies, and generic forms? Literally, _caesura_ means "cutting." It evokes pause. Space for breath, for detours in modes of multispecies literary representation. If the line—working on the assembly line and writing a certain kind of poetic line—is an orientation that draws literature and the abattoir together, as Joseph Ponthus's autofictional poem essay _On the Line: Notes from a Factory_ (2021) suggests, this essay also suggests that the slash is an allied critical-creative orientation that equally requires engagement.

Keywords slash, factory farm fictions, hell, katabatic imaginary, poetic outrageousness

This essay is a broken elemental thing composed of cuts, by which I mean outtakes. Outtakes are scenes or sequences that never make it into a film. The scenes collected here have been retrieved from the cutting floor of my mind's editing suite and reassembled in ways that hold onto an ambitious claim—to think of narrative cuts and silences as interruptive forces in the

Cultural Politics, Volume 19, Issue 1, © 2023 Duke University Press
DOI: 10.1215/17432197-10232544

operation of writing and the imaginative rendering of the abattoir.

Working with outtakes helps me approach, in a new way (for me), questions I have been exploring for years: How can writers critically respond to the existence of abattoirs? What strategies might writers engage to render normalized forms of violence against animals strange and even intolerable through particularly literary practices, strategies, and generic forms?

Literally, *caesura* means "cutting." It evokes a severance and a pause. Space for breath. And in that breath, space for detours of thought.

If the line—working on the assembly line and writing a certain kind of poetic line—is an orientation that draws literature and the abattoir together as is suggested by Joseph Ponthus in his autofictional poem essay *On the Line: Notes from a Factory* (2021), in this essay I suggest that the slash, the cut, the space for the breaths taken between words, is an allied critical-creative orientation that equally requires engagement.

.....

A cut, a slash, a slice, dichotomy—all belong to the abattoir, quantum theory, theories of oppression and justice, and certain pain-filled modes of writing.

The slash is a symbol of hierarchical values (Plumwood 1993). It points to the cuts made in the abattoir from the presticker to the trimmer (Pachirat 2011). It is also a rupture and can be used in writing to tear at customary ways of seeing via "poetic outrageousness," as exemplified by many writings by the Marquis de Sade, including, "It has been estimated that more than fifty million individuals have lost their lives to wars and religious massacres. Is there even one among them worth the blood of a single bird?" (LeBrun 2008: 62).

This, quoted in Annie LeBrun's *The Reality Overload: The Modern World's Assault on the Imaginal Realm*, is used by LeBrun to exemplify poetry's ability to suddenly render present vast plains of violence and poetry's potential to refuse the enshrinement of the status quo, in which the enormity of violence done to more-than-human animals moving through industrial systems is invisibilised. By inverting a horizon of importance, poetic outrageousness reroutes certainties and habits of mind and asks questions that tear at humanist hierarchies.

Poetic outrageousness is bellicose and, I think, belongs to what Maggie Nelson (2011: 11) refers to as the art of cruelty that offers insights into certain "styles of imprisonment." The point is not to uplift or help us forget the violence of the industrialized status quo, but to point to power and violence and to cut us off, momentarily, from everything that feels solid or good. Poetic outrageousness, the art of cruelty, does not alleviate pain but actually is said to contribute to it. Serious consideration of those works might take us somewhere different entirely.

.....

In poetry, a caesura, Latin for "cutting," is a metrical pause or a break in verse where one phrase ends and another begins. It can be expressed by a comma or two lines // slashes.

The slash acts as a recognition of interconnection and entanglements, a suture (Barad 2007, 2010). As Karen Barad (2007) writes, the slash cleaves. And in *Landfill: Notes on Gull Watching and Trash Picking in the Anthropocene*, Tim Dee agrees. He writes that *to cleave* is a verb that "all taxonomists must fear because it means to split and lump, to pull apart and bring together" (Dee 2018: 18).

Cleave violates a metaphysical commitment to the ideas of separateness. It signifies the downfall of individualism and the presumed disconnection of intelligibility from materiality.

The slash is a connective opening, and I use it here to designate the mouth of a passage leading to subterranean places, containers, in which certain lives, stories, deaths are shoved and must find a way to reside.

In these containers I search for things that are the subject of an energetic amnesia in life and so many literary contexts—the lives and deaths of farmed animals.

The passage down isn't straightforward. Its walls are lined with corpses and knives.

One name for the subterranean place is hell. This will not be a heroic descent.

.....

Hell is a slaughterhouse, and I'm not speaking in metaphors. What I mean is that through the slaughterhouse, a certain hell for animals is made immanent in history.

The abattoir is a space divided into finely regulated slices of horror. Like the Chapman brothers' installation artwork *Hell* (1999), made of nine vitrines filled with tiny, defiling toy soldiers chopped up and remodeled, all doing the same thing over and over again, sixty thousand times—killing. Only the soldiers of the abattoir wear red hard hats, yellow hard hats, or white hard hats, according to their line of work.

Unlike the concentric circles of Dante's *Inferno*, the abattoir insists on a strict and amplified linearity. Live animals enter, meat exits.

Detours, swerves, backtracks, *détournements* are forbidden in the abattoir. Detouring equals the transgressive rephrasing of conventional discourse. It is precisely for this reason that I take them as a productive way to write about abattoirs and literature.

This awkward, bloody conceptual ground has been my latitude since 2014, when I became obsessed with thinking and writing about writing—its processes, ethics, and effects—and its relationship to those who are killed on industrial disassembly lines.

.....

The precise space of these containers that you are reading—call them paragraphs, vignettes, fragments, notes—is the obscene, the cultural equivalent to being "off stage" (Chambers 2004). Literary theorist and thinker of AIDS and cultural hauntings Ross Chambers (2004) writes that what is obscene is part of the occult, what is occulturated from society. What is known but not acknowledged, secret, hidden, or covered over, not apprehended by the mind.

Hel means "cover" in Old Norse, a term that attempts to cover horrors that cannot be spoken.

.....

Jenn Ashworth's (2019) memoir *Notes Made while Falling* uses the slash to designate the beginning of an untellable story. "A beginning," she writes, "is a cut in the onward flow of things" (1).

The slash is used to show the intertwining of bodies and worlds, the proliferation of sickening bodies. And it signifies the place where words, sentences, and phrases have dropped off the page, go missing in action. It holds the place of erasure, absence. It shows writing as a leaky container, filled with holes.

Ashworth's entire memoir consists of drawing significant philosophical, aesthetic, and ethical connections between different kinds of cuts that bleed into each other: Alice's rabbit hole is also a wound in the chest that a fist can disappear into is also a hole in the head made by trepanning—a surgical intervention in which a hole is drilled or scraped into a human skull. A slash, cut, or hole in the head is, for Ashworth, a way into narratives of trauma.

Her deep need to explore cuts originates with complications of her C-section delivery, which meant surgeons had to rush her back onto the table and open up her wound. It was during the second operation that her epidural started to wear off. Paralyzed but awake and panicked, she felt unspeakable things: an extreme pressure inside her body, a wind tunnel blowing through her organs. Her legs were rubber. They were dream legs. She needed to move these dream legs to convince the surgeons that she could feel what was going on down below. Speaking of her trauma requires dwelling with*in* and beneath her cuts.

Lori Gruen's (2019) essay "Just Say No to Lobotomy" links the surgical procedure—which also involves a drilling into the skull and severs nerve fibers in the brain that connect the frontal lobe to other brain regions—to forms of affective dismembering wherein talk of love is ablated from discourses on justice for multispecies others and emotions are cut off from cognition.

From lobotomy, an act of severance, Gruen talks of entangled empathy. Gruen, like Ashworth, like me, is anchored by the many possibilities of the slash to trace certain violences and write/imagine an otherwise.

.....

Regard the pain of animal others. The gun must be placed perpendicular on the skull surface, at an imagined center point. The X.

Imagine two lines drawn from the tops of the eyes intersecting in the middle of the skull. This guide is only for certain cows. A Holstein dairy cow, which has a long head, requires a different imagining. Their physiology is different, specific, unique.

When you aim for the X you are actually aiming for the hippocampus, the point of crossover for the nerve fibers of both hemispheres of the brain.

The idea of the bolt is to render a cow insensible. Induce a percussive wave within the brain tissue, increase the disruption to nervous impulses, which result in a *shearing* causing physical damage to the cells.

How to determine insensibility?

The head and neck must be loose and floppy—rag-like.

The tongue should hang out—straight and limp.

(A tongue curling in and out of the mouth *may* be a sign of sensibility.)

Eyes should be wide.

The stare blank.

There should be no blinking or corneal reflex in response to touch.

The effectiveness of this operation comes down to several things: the maintenance of equipment, experience of operator, correct settings, and use of traumatizing instruments.

Inevitably there is the very real possibility of poor gun placement.

Inevitably there is the very real possibility of a cow remaining conscious until they are bled out (Grandin 2017).

These puncture wounds are a manifestation of hellholes.

While I write these notes I am also writing a book about the hell of writing a scholarly monograph about the hell of industrial animal agriculture.

So, two types of hell are involved in the meta framing of the book. One is of the brutalism of systems of animal confinement, killing, and meat production. The other is of trying to speak this violence in an environment uninterested in hearing about it.

Forms of hell accumulate but do not approximate in this process.

After years of not knowing how to begin the book, I realize this hole is exactly the hole that must bring me into my own funereal story-mind. It must act as a severance and a reverse severance cleaving me to the life and death of farmed animals.

.....

When I first learn about caesuras, their use in poetry, I repeatedly make the mistake of writing that the slash signals a juncture between worlds. *The caesura is the point at which one world ends*, I wrote in an essay for my poetry class, *and the following begins*. I was marked down for having written *world* instead of *word*. But the mistake feels truer, bigger, heavier, more consequential than its corrective.

In music, a caesura is a pause, a silence, time not counted, and represented on sheet music by two slashes sometimes referred to as train tracks.

Détournement is a strategy for derailing, wandering off stage, getting off topic. It is figurative roller-coastering.

Rachel Falconer (2007), literary scholar of hell on earth, notes that in contemporary writing hell is treated as a caesura, an infernal pause. A severance. Like what Russian antiformalist scholar Mikhail Bakhtin (1981) called a chronotope—a generically distinct representation of space and time in narrative.

.....

A piggery is a suspended place and time, a hellhole filled with beings turned into nonentities, ghosts. "An animal is an animal, and a pig is not even that" (Del Amo 2019: 235).

In 2019 the English translation of Jean-Baptiste Del Amo's novel *Animalia* was published. A friend suggested I read it. *It's disgusting*, she said and wished me luck.

The book is praised as a modern-day classic for its commitment to all that is stained, spoiled, and violated (Sansom 2019).

I started with gusto but soon could only read through parted fingers. At first, I placed this book on a literary shelf alongside the Marquis de Sade for its fetishization of mastery over flesh and consumption, its dirty, flayed, defiled, and scarified bodies—but mostly for its insistence that to live as an animal is to suffer.

But then I realized that there is no fetishization of violence in these pages, only a faithful—forensic—rendering of the endless work of killing for a living. A faithful rendering of a hellhole in which hordes of pigs are trampled, bitten, beaten, shot with bolt guns, stuck through the thigh with a hook and hoisted up to bleed out, drop dead (brought down by heart attacks).

The line between a fetishized and forensic account of abattoir as hell on earth is finely sliced. The novel begins in a rural hell. It etches out the morose story of a life lived on a pig farm in a fictional French village—Puy-Larroque—and walks down a muddy-bloody narrative track, following five generations of a family from one century to another, one tradition to another—peasant farm to industrial piggery—on a trajectory of descent.

It goes deep. Walks well into the mental night that turns factory farms, concentrated animal feeding operations, finishing sheds, and abattoirs into "shadow places"—disregarded places that elude responsibility (Plumwood 2008).

Not a narrative of technological or human progress and development but the brutality of industrialism with its slurry-pits and chemical infestations (Lindane, DDT [Dichlorodiphenyltrichloroethane], Chlordane, and PCBs [Polychlorinated biphenyls]), which deteriorate all life pulled into their orbit.

The industrial piggery at the center of the narrative—renovated, modernized, endlessly expanding—"cannot contain what it needs to constantly assimilate and regurgitate" (Del Amo 2019: 238) And what passes between the covers is intensifying madness. Tides of shit and prolapsed vaginas. Vaccines, hormones, pesticides—all poisons. Pigs lie in excrement to try cool down in the humidity and rancidity of their sheds.

Modified complexly for human desires, born to grow almost all muscle and no fat, the sickly pigs live 182 days spent "vegetating in the half-light of a pig unit" (258). And death is hard, of course. Short life, hard death. In this fictional (but so very truth-filled) universe violence is everlasting and porcine injustice is infinite.

One thought dominates me while I read. Del Amo kept his mind in this hell for the years it took to write the book. I can't be sure of this, of course. But that's what is pressed onto me as I read this katabatic work.

The "katabatic imagination" is a worldview that conceives of selfhood as a narrative construction of an infernal quest or journey, and a return.

Literally, *katabasis* means "going down" and is most notably associated with Dantean or Christian hinge narratives of descent into an underworld and a return to the overland.

Classical katabasis offers a narrative of heroic self-discovery, knowledge gathering, and self-realization, and most frequently it associates the quester as a man who descends into the female earth/underworld (Falconer 2007: 7). Historically, Rachel Falconer writes, hell in literature has associations with madness and femininity.

As Falconer explores, 1945—the midway of a century overrun with wars, genocides, systemic injustices of multiplying kinds, rapacious exploitations of the earth, and environmental catastrophe—ushered in literature that spoke of hell on earth, which was embodied in gulags, prisons, the aftermath of bomb blasts, and forms of environmental poisoning. To this list I would like to add the post–World War II intensification of factory farming, with its amplified production orientations that have led to the vertical integration of farms—meaning a company has control of an animal's body from birth until death, often in the same facility—the introduction of artificial growth hormones in feed and the use of antibiotics. This postwar intensification represents a time when animal subjects were living, moving through, and dying from an infernal journey in previously unmatched numbers and levels of suffering.

Animals are not meant to return from the hell of the factory farm. Can a writer?

.....

The word *obscene* first appeared to me in J. M. Coetzee's (1999) novel *Elizabeth Costello* when the central character, a novelist called Elizabeth Costello, meditates on the ethical relationship between unspeakable acts of violence and attempts made by writers to speak them, to body them into the world.

The part of the book that I want to point to is when Costello, preparing to give a lecture at a university in Amsterdam, questions whether telling stories is a moral good in itself. She has become convinced that people aren't always improved by what they read. What is being risked, she asks herself, when a writer walks into dense forests of violence to research and tell or retell certain horrific histories? There is no guarantee that the soul will return unharmed. What she imagines of the writer's movement resembles the narratives of descents into an underworld.

When Costello tries to reconcile the desire for writers to enter the bloody basements of history and represent them to the world, she imagines a quiet procession of soul-damaging stories passing from person to person. If humanity is to be saved from itself, she thinks, some stories must be kept off the cultural stage, held in the obscene.

.....

TRUTH, reads the sign. Beside it, a woman in a Guy Fawkes mask carries a TV screen with footage of pig being beaten down a chute toward a knocker—which is a person who uses an air gun to drive a captive-steel bolt into a pig's forehead.

This woman has been holding the device as an offering of truth-telling and witnessing for about an hour. I've been standing there watching the footage loop for almost the same amount of time. The repetition without variation is brain-numbing.

A man is hitting a pig. A pig is crouching. A pig is squealing. A pig runs toward the knocker. A pig is being hit. A man is hitting a pig. A pig is crouching. And on and on . . .

This is the story of the world. It matters how this story is told.

At the protest, people walk past, some glance sideways, others not at all. A woman yanks her son to the other side of the street. A man stops and watches for a few moments. Soon he looks around to see if other people are watching him watch the man beat this pig. He says loudly, to the TV—and also, I think, to the world—*WELL THIS IS JUST DISAPPOINTING*. Then he, too, walks away.

His statement hits me with its willful passivity, filled with absent possibility that anything in this footage will ever change. (The man is able to walk away at the moment of his choosing without any threat of danger or pain.)

Maybe he'll go home, watch a movie, and forget all about it. Or maybe he'll clear his kitchen of meat, eggs, and dairy and bury them in the backyard. Either way, his outburst recalls what Namwali Serpell (2019) writes in "The Banality of Empathy"—narrative might simulate empathy, but that doesn't mean it *stimulates* it.

In this essay, Serpell takes apart the cultural orthodoxy that the ethics of the novel lies in its ability to bring the one to the many and the many to the one and in doing so will give rise to social good. Serpell traces this idea back to the eighteenth century with the writer George Eliot, who in a letter to Charles and Cara Bray said, "if Art does not enlarge men's sympathies, it does nothing morally" (quoted in Serpell 2019). Important to note here: at the time, *sympathies* meant something closer to what we call empathy now. For Serpell the equation empathetic art = better world doesn't stack up.

Seeing real violence doesn't guarantee ethical action, so why, she asks, "do we think art about suffering will?" Actually, Serpell suggests that the empathy model can work, perversely, as a "gateway drug" to saviorism, which triggers emotional experiences while counteracting any actual ethical action. Thus, oppressive practices are preserved.

.....

Neither *Animalia* nor the protest documentary simulates empathy, though I think they offer a representation of hell for animals that can invite a reader to turn their mind to the position of an/other. But this isn't necessarily a form of imaginative dark tourism, nor does it have to equate to assimilation of self and other.

This kind of imagining is less about moral feelings than it is about political justice, "offering a broader view of humanity, while maintaining a keen awareness of who is friend and who is foe" (Serpell 2019). So Serpell moves beyond the idea of writers imagining their way into the life of another because there is a long history of an instrumentalizing and violent literary "empathy." "Writers," she says, "are not historically renowned for being good people."

What Serpell is advocating is the expansion of the scope of representations through the question: Who is represented and how? A model for ethical writing is, she suggests, to imaginatively inhabit a *position*, not a *person*. And, importantly, this is not to be done just anywhere, but rather where you are *welcome*.

.....

Working through complexities of subjectivity and point of view in relation to beings-other-than-human is deeply significant to discussions of empathy, literature. and multispecies justice, particularly in the fields of critical animal studies, literary animal studies, and what might be called multispecies literary justice (Nagel 1974; Coetzee 2004; Gruen 2015).

Multispecies literary justice does not aim at individual moral improvement through reading/writing—the idea that "art encourages empathy and empathy will save us all" (Serpell 2019), nor that art enacts a kind of forceful, "orthopaedic" intervention on society (Nelson 2011: 4). It moves toward the idea that multispecies literature broadens representations of multispecies life and death and, doing so, can imaginatively build solidarities across differences, similarities, and complexities of beings. I think about it now as a form of literary political representation that teaches writers to re-cognize the world as it is, to level a rupture in default stories that mask the world's material realities and violences.

And more, it invites you give up on the familiar—"voracious," as Serpell writes—practice of inhabiting others and instead to imaginatively visit with them, to ask the question: What would I want the world to do for me if I was born into the situation of meat?

The project is enormous and, I think, incompletable. *But* it is related to one of the most pressing questions facing writers in the twenty-first century—*How do we reduce the violence, hatred, and deadly indifference that have so often marked human and more-than-human animal interactions?*

This question considers how to lessen the amount of cruelty in the world. I want to think about *that*, yes. Though I am stuck working through expressions of cruelty and representations of hell on earth for animals and asking what they might offer a world already congested by these very things.

Thinking back to Elizabeth Costello, I have to ask, again: What is the ethical wager for the writer who brings to life (once more) the details of even one of these cruelties, let alone their intersectional sadisms?

What is the ethical risk a writer

takes when she trespasses on the life of another's death? Into whose death can she be welcomed? Like Maggie Nelson (2011) in *The Art of Cruelty: A Reckoning*, I would like to understand more about compassion, and so I take a risk she has already taken and I study cruelty, carefully.

Literature, writes James Wood (2008: 52), is an art that teaches one to notice. While life is full of amorphous detail, literature directs one's focus to certain details, helps develop the ability to recognize the most ordinary thing as a most remarkable thing. What is merciless is also banal, what is horrifying is everyday, and what has become everyday is actually horrifying. For Woods, detail brings the fullness of living, a "thisness"—specificity of detail that dissolves abstractions with sheer palpability—to a work of fiction. Specificity is a kind of cruelty that can be enacted in writing.

Earlier, I said *Animalia* wasn't a Sadeian novel. But that's not exactly right. What *is* Sadeian about it is its refusal to allow knowledge of cruelty to remain general. Part of the novel's cruelty is its commitment to specificity, its commitment to realism. I don't want to be general about realism, it's too important, so let me say that I'm talking industrial realism and realism of the flesh.

Industrial realism—individuals and masses represented at work in factories and sheds; intimate moments within huge labor-filled environments; aesthetic exploration of sweaty bodies, exhausted lunch breaks, blood, and injury; everyday existence. To show a truthfulness about the laboring world through art.

Here is a slice of flesh realism in action from Del Amo (2019: 258).

> Because everything in the closed, stinking world of pig-rearing is simply one vast infection, constantly contained and controlled by men, even the carcasses churned out by the abattoirs to stock the supermarkets, even when they have been washed with bleach, cut into pink slices and packed in cellophane into pristine white polystyrene trays, they bear the invisible taint of the pig shed, minute traces of shit, germs and bacteria, against which the men fight a losing battle with their puny weapons: high-pressure hoses, Cresyl, disinfectants for the sows, disinfectant for wounds, worming pellets, vaccines for swine flu, vaccines for parvovirus, vaccines for Porcine Reproductive & Respiratory Syndrome, vaccines against porcine circovirus, iron injections, antibiotic injections, vitamin injections, mineral injections, growth hormone injections, food supplements—all this in order to compensate for deficiencies deliberately created by man.

Industrial and flesh realisms are rendered through a commitment to detail. Detail is also part of their commitment to exploring a form of representative thinking that is openly, radically incomplete, a wavering, momentary suffering shared between subject, writer, and reader.

Critically, it is shared *at a distance*, does not pretend to collapse the distance between subject, writer, and reader. Nor should it be thought to exist on a scale analogous to the suffering of those pigs who physically languish under the dim lights of a shed and miscarry onto their own feces. Seven dead piglets in one stall, nine in another. All to go in the incinerator.

.....

When I think of Elizabeth Costello's ethical knots I ask myself, If enough cruel stories build up in our systems, will they do damage to the living tissue of the world? I think Barad would say yes, since the dichotomy discourse/materiality is false, is dis/continuous. Words and things, speech and action are always threading through one another—waves entwining as they drive into the shore.

And there are these words from Hannah Arendt (1969: 80) in *On Violence*, "The practice of violence, like all action, changes the world, but the most probable change is to a more violent world."

The representation of violence through poetic outrageousness and more rigorous imaginative modes of writing can bring clarity to the world as it is. As I see it, it is a matter of loyalty to the dead to re-presence them among the living. This is not a conclusion, but it is what I am sitting with now.

.....

Back on the street, in front of the protest, there's this guy who enters the scene. He moves to stand in front of the TV that the woman in the Guy Fawkes mask is still holding. The man looks at the crowd and performatively unwraps a McDonald's burger. He stuffs it in his mouth, eats with over-the-top orgasmic delight—eyes rolling, pelvis grinding. I raise my phone to take a photo. He gives me a sauce-wet smile and a big thumbs up.

He is a man standing at the mouth of hell.

References

Arendt, Hannah. 1969. *On Violence*. New York: Harcourt.

Ashworth, Jenn. 2019. *Notes Made while Falling*. London: Goldsmiths.

Bakhtin, Mikhail. 1981. *The Dialogic Imagination*. Austin: University of Texas Press.

Barad, Karen. 2007. *Meeting the Universe Halfway: Quantum Physics and the Entanglement of Matter and Meaning*. Durham, NC: Duke University Press.

Barad, Karen. 2010. "Quantum Entanglements and Hauntological Relations of Inheritance: Dis/continuities, SpaceTime Enfoldings, and Justice-to-Come." *Derrida Today* 3, no. 2: 240–68.

Chambers, Ross. 2004. *Untimely Interventions: AIDS Writing, Testimonial, and the Rhetoric of Haunting*. Ann Arbor: University of Michigan Press.

Coetzee, J. M. 2004. *Elizabeth Costello: Eight Lessons*. London: Vintage.

Dee, Tim. 2018. *Landfill: Notes on Gull Watching and Trash Picking in the Anthropocene*. White River Junction, VT: Chelsea Green.

Del Amo, Jean-Baptiste. 2019. *Animalia*. Translated by Frank Wynne. Melbourne: Text.

Falconer, Rachel. 2007. *Hell in Contemporary Literature: Western Descent Narratives since 1945*. Edinburgh: Edinburgh University Press.

Grandin, Temple. 2017. "Recommended Captive Bolt Stunning Techniques for Cattle." Department of Animal Science, Colorado State University. https://www.grandin.com/humane/cap.bolt.tips.html.

Gruen, Lori. 2015. *Entangled Empathy: An Alternative Ethic for Our Relationships with Animals*. New York: Lantern.

Gruen, Lori. 2019. "Just Say No to Lobotomy." In *Animaladies: Gender, Animals, and Madness*, edited by Lori Gruen and Fiona Probyn-Rapsey, 11–24. New York: Bloomsbury Academic.

LeBrun, Annie. 2008. *The Reality Overload: The Modern World's Assault on the Imaginal Realm*. Translated by Jon E. Graham. Rochester, NY: Inner Traditions.

Nagel, Thomas. 1974. "What Is It Like to Be a Bat?" *Philosophical Review* 83, no. 4: 435–50.

Nelson, Maggie. 2011. *The Art of Cruelty: A Reckoning*. New York: W. W. Norton.

Pachirat, Timothy. 2011. *Every Twelve Seconds: Industralized Slaughter and the Politics of Sight*. New Haven, CT: Yale University Press.

Plumwood, Val. 1993. *Feminism and the Mastery of Nature*. New York: Routledge.

Plumwood, Val. 2008. "Shadow Places and the Politics of Dwelling." *Australian Humanities Review*, no. 44. http://australianhumanitiesreview.org/2008/03/01/shadow-places-and-the-politics-of-dwelling/.

Ponthus, Joseph. 2021. *On the Line: Notes from a Factory.* Translated by Stephanie Smee. Melbourne: Black Inc.

Sansom, Ian. 2019. "War, Violence, Sickness, and Cruelty." Review of *Animalia*, by Jean-Baptiste Del Amo. *Guardian*, May 30. https://www.theguardian.com/books/2019/may/30/animalia-jean-baptiste-del-amo-review.

Serpell, Namwali. 2019. "The Banality of Empathy." *New York Review of Books*, March 2. https://www.nybooks.com/daily/2019/03/02/the-banality-of-empathy/.

Wood, James. 2008. *How Fiction Works*. London: Jonathan Cape.

Hayley Singer is precariously employed as a lecturer in creative writing at the University of Melbourne. Her research and writing practice moves across the fields of creative writing, ecofeminism, and critical animal studies. Her first book is *Abandon Every Hope: Essays for the Dead* (2023).